Science, Technology and Medicine in Modern History

General Editor: **John V. Pickstone**, Centre for the History of Science, Technology and Medicine, University of Manchester, England (www.manchester.ac.uk/CHSTM)

One purpose of historical writing is to illuminate the present. At the start of the third millennium, science, technology and medicine are enormously important, yet their development is little studied.

The reasons for this failure are as obvious as they are regrettable. Education in many countries, not least in Britain, draws deep divisions between the sciences and the humanities. Men and women who have been trained in science have too often been trained away from history, or from any sustained reflection on how societies work. Those educated in historical or social studies have usually learned so little of science that they remain thereafter suspicious, overawed, or both.

Such a diagnosis is by no means novel, nor is it particularly original to suggest that good historical studies of science may be peculiarly important for understanding our present. Indeed this series could be seen as extending research undertaken over the last half-century. But much of that work has treated science, technology and medicine separately; this series aims to draw them together, partly because the three activities have become ever more entwined. This breadth of focus and the stress on the relationships of knowledge and practice are particularly appropriate in a series which will concentrate on modern history and on industrial societies. Furthermore, while much of the existing historical scholarship is on American topics, this series aims to be international, encouraging studies on European material. The intention is to present science, technology and medicine as aspects of modern culture, analysing their economic, social and political aspects, but not neglecting the expert content which tends to distance them from other aspects of history. The books will investigate the uses and consequences of technical knowledge, and how it was shaped within particular economic, social and political structures.

Such analyses should contribute to discussions of present dilemmas and to assessments of policy. 'Science' no longer appears to us as a triumphant agent of Enlightenment, breaking the shackles of tradition, enabling command over nature. But neither is it to be seen as merely oppressive and dangerous. Judgement requires information and careful analysis, just as intelligent policy-making requires a community of discourse between men and women trained in technical specialities and those who are not.

This series is intended to supply analysis and to stimulate debate. Opinions will vary between authors; we claim only that the books are based on searching historical study of topics which are important, not least because they cut across conventional academic boundaries. They should appeal not just to historians, nor just to scientists, engineers and doctors, but to all who share the view that science, technology and medicine are far too important to be left out of history.

Titles include:

Julie Anderson, Francis Neary and John V. Pickstone
SURGEONS, MANUFACTURERS AND PATIENTS
A Transatlantic History of Total Hip Replacement

Roberta E. Bivins
ACUPUNCTURE, EXPERTISE AND CROSS-CULTURAL MEDICINE

Roger Cooter
SURGERY AND SOCIETY IN PEACE AND WAR
Orthopaedics and the Organization of Modern Medicine, 1880–1948

David Edgerton
ENGLAND AND THE AEROPLANE
An Essay on a Militant and Technological Nation

Jean-Paul Gaudillière and Ilana Löwy (*editors*)
THE INVISIBLE INDUSTRIALIST
Manufacture and the Construction of Scientific Knowledge

Thomas Schlich
SURGERY, SCIENCE AND INDUSTRY
A Revolution in Fracture Care, 1950s–1990s

Eve Seguin (*editor*)
INFECTIOUS PROCESSES
Knowledge, Discourse and the Politics of Prions

Crosbie Smith and Jon Agar (*editors*)
MAKING SPACE FOR SCIENCE
Territorial Themes in the Shaping of Knowledge

Stephanie J. Snow
OPERATIONS WITHOUT PAIN
The Practice and Science of Anaesthesia in Victorian Britain

Carsten Timmermann and Julie Anderson (*editors*)
DEVICES AND DESIGNS
Medical Technologies in Historical Perspective

Science, Technology and Medicine in Modern History
Series Standing Order ISBN 978–0–333–71492–8 hardcover
Series Standing Order ISBN 978–0–333–80340–0 paperback
(*outside North America only*)

You can receive future titles in this series as they are published by placing a standing order. Please contact your bookseller or, in case of difficulty, write to us at the address below with your name and address, the title of the series and the ISBNs quoted above.

Customer Services Department, Macmillan Distribution Ltd, Houndmills, Basingstoke, Hampshire RG21 6XS, England

---

Surgeons, Manufacturers and Patients

A Transatlantic History of Total Hip Replacement

Julie Anderson, Francis Neary and John V. Pickstone
Centre for the History of Science, Technology and Medicine
The University of Manchester

In collaboration with James Raftery
Wessex Institute for Health Research and Development
The University of Southampton

palgrave
macmillan

First published 2007 by
PALGRAVE MACMILLAN
Houndmills, Basingstoke, Hampshire RG21 6XS and
175 Fifth Avenue, New York, N.Y. 10010
Companies and representatives throughout the world

PALGRAVE MACMILLAN is the global academic imprint of the Palgrave Macmillan division of St. Martin's Press, LLC and of Palgrave Macmillan Ltd. Macmillan® is a registered trademark in the United States, United Kingdom and other countries. Palgrave is a registered trademark in the European Union and other countries.

ISBN 13: 978–0–230–55314–9 hardback
ISBN 10: 0–230–55314–1 hardback

This book is printed on paper suitable for recycling and made from fully managed and sustained forest sources. Logging, pulping and manufacturing processes are expected to conform to the environmental regulations of the country of origin.

A catalogue record for this book is available from the British Library.

A catalogue record for this book is available from the Library of Congress.

10 9 8 7 6 5 4 3 2 1
16 15 14 13 12 11 10 09 08 07

Printed and bound in Great Britain by
Antony Rowe Ltd, Chippenham and Eastbourne

In Memoriam

Florence Pickstone
(1913–2001)

Helen Falconer Anderson
(1915–2002)

Dorothy Neary
(1922–1992)

Contents

List of Illustrations

List of Tables and Figures

Tables

Figures

Acknowledgements

This book is a collective effort and we are named alphabetically. John Pickstone supervised the project, focused on medical innovation and health services and is largely responsible for the final text. Julie Anderson was the main researcher and prepared most of the initial drafts, especially on Thackray, the American history, patients, and materials. Francis Neary worked on the project for a year, and continued to input from his employment at Wrightington Hospital, especially on the technologies and the British surgeons. We have enjoyed the participation of James Raftery, for the quantitative data and stimulating discussion of their meaning. But the book has many other contributors, whom we can now sincerely thank for invaluable assistance over a long gestation.

The Economic and Social Research Council (ESRC) provided the funding for this project as part of the Innovative Health Technologies (IHT) research programme. David Cantor helped draft the initial application, before he went to work in the US. Andrew Webster led the whole IHT programme admirably and Stephanie Gant provided administrative support. We were grateful for the opportunity to present our work at IHT meetings and for the feedback from other projects in the programme. Our colleagues at the University of Manchester, Andrew James and Stan Metcalfe, working on the IHT project on the intra-ocular lens, were particularly helpful.

Michael Worboys, as Director of our Centre for the History of Science, Technology and Medicine (CHSTM), extended the research funding, and we also much appreciate his personal encouragement and support. Much of the other work of the Centre is supported by the Wellcome Trust, to whom we are deeply grateful.

We would like to thank the John Charnley Trust for allowing Francis to continue researching on the project, while supporting his museum projects. Christopher Faux, Douglas MacMillan, Lady Jill Charnley and Tristram Charnley were instrumental in this respect. Tristram and Lady Charnley also provided us with a fascinating and personal insight into Sir John's development of surgical techniques and his early relationships in the medical devices industry. Chris Faux also read earlier versions of the text and was a great supporter of the project.

Many surgeons assisted our research. In the UK, Kevin Hardinge and the late Robin Denham provided a wealth of information, and at

Wrightington, Mike Wroblewski, Bodo Purbach and Martyn Porter were always on hand to answer difficult questions. Elsewhere, John Kirkup, Robin Ling and John Older were generous with their time. Claire Darrah, Keith Tucker and the late Hugh Philips gave Francis insights into McKee's work and the Norwich THR programme, as did Peter Ring and Kaild Drabu for Redhill.

In the US, Julie would like to particularly thank Wayne Paprosky at Rush Presbyterian and Central Dupage Hospital in Chicago for allowing her to disrupt his clinic for a few days in September 2002. She is also grateful to Allan Gross, for the queries answered and the doors opened. At Rush Presbyterian in Chicago, Craig Della Valla, Jorge Galante and Mitchell Sheinkop gave valuable time, as did William Harris at the Massachussets General Hospital (MGH) MGH in Boston and Don Howie in Australia. The American Academy of Orthopaedic Surgeons (AAOS) allowed access to surgical courses, their meetings and their extensive collection of hip prostheses. She would particularly like to thank Bill Tipton and Lynne Dowling from AAOS. We are also grateful to the many surgeons whose papers we attended at meetings of the British Hip Society, British Orthopaedic Association and Low Friction Society in the UK and the AAOS in the US.

Biomedical engineers were also very helpful. The late Harry Craven gave up his time and was immensely patient with requests for further details and information. Denis Bobyn gave insights into research and development, as did Paul Potsak and his colleagues from St Louis, who also entertained Julie at the AAOS meeting in New Orleans. Clive Lee and the late John Scales made time to speak to Francis.

While working on the new patient information centre and museum at Wrightington Hospital, Francis was fortunate to discuss ideas with well-informed and generous colleagues, including the current Wrightington engineers, Brian Derbyshire and Frank Brown. The former engineers, Jeff Middleton and the late Hugh Howorth, came in to demonstrate some of Charnley's equipment and techniques. Peter Kay and Tom Pickering were very helpful in discussing recent strategic initiatives within the NHS.

Of the manufacturers, Julie would also like to thank Dane A. Miller and his team at Biomet, who allowed her access to their factory and were truly friendly. She thanks Tom Allen, Craig Blaschke, David Brown, Tony Fleming, Barb Goslee, Randy Hulse, John Sasuraba, Dave Schroder, John Waggoner and Lonnie Witham for giving up their time; and Phil Gibbs and Ron Gilbert who spent hours explaining designs and markets. Both Dane Miller and his wife Mary Louise were wonderfully hospitable on non-working days. At Zimmer, Roy Crowninshield was very helpful, as was Brad Bishop who gave a tour of Zimmer's historical display.

Many patients talked of their experiences, in clinics and phone calls; and relatives also helped us see the hopes attached to surgery. They remain anonymous, for the most part, but they have our thanks.

This book would have been much poorer without the assistance of our colleagues from museums and libraries. Alan Humphries at the Thackray Museum in Leeds and Simon Chaplin at the Royal College of Surgeons were immensely supportive. Simon also provided space for an exhibition he organised with Francis. At the John Rylands Library at the University of Manchester, we would like to acknowledge the constant support and assistance of both Stella Butler and James Peters. The surgeon Ravi Kunzru suggested the Wellcome Witness seminar, which Tilli Tansey, Daphne Christie, Lois Reynolds, Alan Lettin and Francis helped organise. We also thank the various seminar and conference audiences to whom we presented parts of this work.

Several colleagues outside CHSTM have been especially informative and supportive, and have commented on early drafts to good effect. We thank Pablo Gómez, orthopaedic surgeon turned historian, and Rajan Madhok, a surgeon turned to public health. Stuart Blume advised on technologies, Stephen Harrison on health policy, and Rosemary Stevens supplied a late dose of sophisticated good sense. Thomas Schlich, at various stages, shared our questions.

Our colleagues in CHSTM helped greatly on a professional and personal level: particularly, Flurin Condrau, Vanessa Heggie, Aya Homei, Vladimir Jankovic, Robert Kirk, Gillian Mawson, Neil Pemberton, James Sumner, Carsten Timmermann, Elizabeth Toon, Simone Turchetti, Helen Valier and Duncan Wilson.

Others, outside CHSTM, also provided background information, read a chapter, listened to 'hip stories', or simply gave support to one or more of us; they include Emma Brennan, Andrew Biswell, Joanna Bourke, David Cantor, Roger Cooter, Ana Carden-Coyne, David Edgerton, Matthew Frost, Jane Henderson, Steve Ickringill, Kara McKechnie, Heather Perry, Ranulph Redlin, Dean Robertson, Lutz Saurteig, Walton Schalick, Chandak Sengoopta and Steve Sturdy.

Seona Owtram, Helen Valier, Neil Pemberton and Diane Miller have been especially important and we are singularly grateful. We each also thank our families, who have also waited long for the replacement of hips as our joint preoccupation.

Finally, we would each dedicate our book to the memory of a loved one for whom THRs were important: to Julie's grandmother, Helen Falconer Anderson, whose THR in Australia gave three precious and pain free years; to John's mother, Florence Pickstone, whose THR at the Victoria Hospital in Burnley, Lancashire, worked well for over two decades; and to Francis' grandmother, Dorothy Neary, whose heart sadly failed as a THR was attempted for a hip fracture caused by a fall at home. Thus do authors, technologies and lives interact.

Notes on Authors

Dr. Julie Anderson is a Research Associate in the Centre for the History of Science, Technology and Medicine (CHSTM) and Wellcome Unit for the History of Medicine at the University of Manchester. Her areas of research include the history of disability and blindness and she has published on disability and gender, sport and medicine. With Carsten Timmermann she edited *Devices and Designs: Medical Technologies in Historical Perspective* (2006). She is currently completing a book on the history of rehabilitation in the Second World War.

Dr. Francis Neary is a Research Associate at CHSTM, University of Manchester. After completing his doctoral thesis on human consciousness and scientific biography in the late 19th century at the University of Lancaster, he worked on the history of psychiatry in the 20th century at the University of Exeter. More recently at CHSTM, he has been researching on the history of material culture and innovation in joint replacement technologies and he has curated three exhibitions on the past, present and future of hip replacement.

Professor John V. Pickstone has worked in Manchester since 1974, and founded the Wellcome Unit and the CHSTM in 1986. He directed the Centre till 2002, since then he has been a Research Professor. His present research is mostly on recent medicine, especially on cancer services, medical technology, and the NHS around Manchester. He is the author of *Ways of Knowing: a New History of Science, Technology and Medicine* (2000), and editor, with Roger Cooter of the *Companion to Medicine in the Twentieth Century* (2002).

Professor James Raftery is Director of the National Coordinating Centre for Health Care Technology Assessment, and of the Wessex Institute for Health Research and Development at the University of Southampton's School of Medicine. He previously directed the Health Economics Facility at the University of Birmingham. His interests include the use of modelling in evaluating healthcare interventions, economic evaluation alongside clinical trials, and the use of routine data in health technology assessment.

1
Total Hip Replacement: Introduction, Sources and Outline

On May 9 1933, an article in *The Times* headed 'Railway Engineer's Suicide', reported that a 57-year old man had shot himself 'while of unsound mind.' As the Coroner explained,

> (He was) Suffering from profound depression caused by physical disability arising from a chronic disease of the hip-joint, which disabled him, which was getting worse, and of which there was no hope of a cure.[1]

The majority of sufferers learned to live with their painful and disabling condition, but that life could be very difficult. As one surgeon wrote of osteoarthritis,

> Here we have a man, otherwise physically capable and in robust health, who is rendered useless and frequently becomes a hanger-on of charitable associations, on which he is dependent for his daily bread, all because of a painful hip joint.[2]

While pain killers such as aspirin could be relied on for a period of time, there was certain to be a point when they ceased to work well, and no other anti-inflammatory drugs were then available.[3]

This book traces the history of the Total Hip Replacement (THR), which is counted as one of the major successes of post WWII surgery. From the 1960s it has provided a reliable means of relieving pain and increasing the mobility of those who suffered from arthritic hips and related disorders. We focus on Britain, where THR was invented and then on the US; we mention other countries, but this is not a global history.

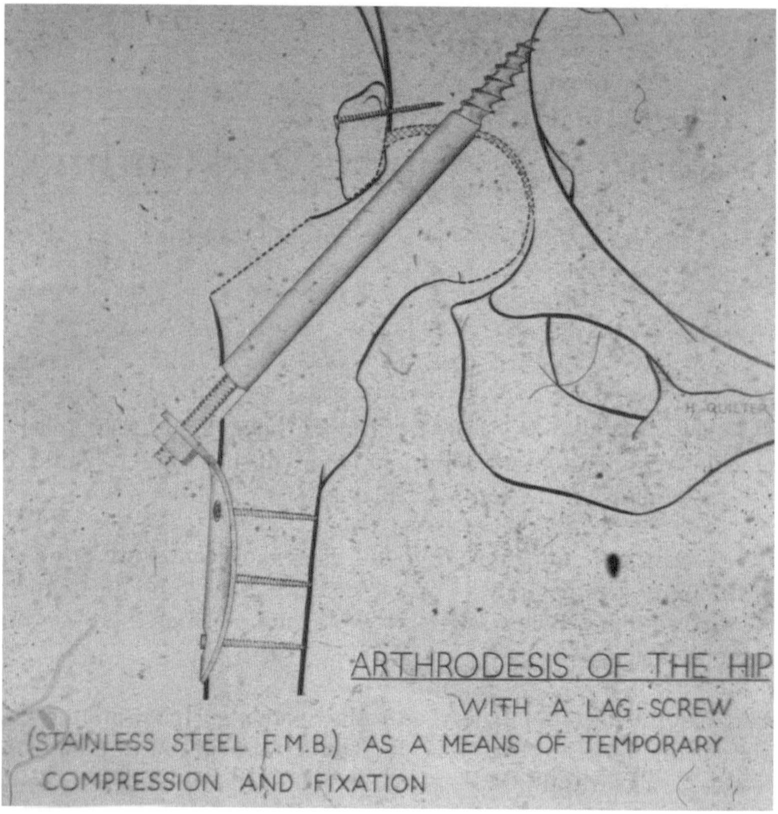

1.1 A Diagram showing a Technique for Arthrodesis of the Hip in the 1930s

Because osteoarthritis was so painful and debilitating, there had been a long history of surgical experimentation, inserting various materials into the joint, or cutting away various parts. The hip joint could be fixed (arthrodesis); or the top of the femur (thigh bone) could be surgically divided to restructure a deformed joint (osteotomy). But these measures were only practicable for one hip, and sooner or later most patients had problems with both. An alternative set of solutions were developed from the 1920s, involving the insertion of artificial parts.

The head of the thigh bone or femur might be replaced with a metal or ceramic head fastened to the stem of the bone; or in some cases a cup of metal was inserted into the hip socket (or acetabulum). These prostheses were mostly developed by American and French surgeons, and by the 1950s, replacing the femoral head was the preferred treatment by some surgeons for certain patients, especially in the US.[4] Orthopaedic companies supplied femoral prostheses, but most surgeons were sceptical as to their efficacy. For orthopaedic surgeons who saw immobilisation and other conservative practices as the most scientific way to promote healing and restore functions, femoral prostheses were one of a series of drastic interventions whose history had been deeply problematic. Prostheses often broke or they became loose or infected.

But the world of surgery changed with WWII. The introduction of antibiotics, the routinisation of blood transfusions, the improvement of anaesthesia and better ways of handling shock, had all encouraged surgeons to be more ambitious. Especially in the US, the war introduced a new wave of heroic surgery, epitomised in the ever more radical operations for breast cancer.[5] In Britain, there was a new Labour government and a new National Health Service (NHS), and the problems facing medicine seemed to have changed: infectious diseases were in retreat and the medical effects of poverty would become history – or so it seemed. The new challenges for medicine would be chronic diseases, which seemed to affect all social classes, and inherited disorders

1.2 Children Suffering from TB in the 1930s, treated by rest and fresh air at Wrightington Hospital

which might eventually be understood and controlled. We can see the shift by looking at tuberculosis (TB) of bones and joints.

TB services epitomised the public medicine of the interwar years; they included health education, domiciliary care, dispensaries, sanatoria, specialists in thoracic surgery and orthopaedic hospitals for the (conservative) treatment of TB of bones and joints.[6] But by the late 1950s, it was clear that TB could be cured by antibiotics, and that new forms of vaccination would prevent it getting hold of the young. Some orthopaedic surgeons, newly salaried under the NHS, looked to new challenges and especially to the relief of crippling diseases such as osteoarthritis which, with rheumatism, had attracted medical and charitable attention largely because of their effects on workers.[7]

It was in postwar Britain that surgeons experimented with prostheses for *both* parts of the hip joint, designing replacements for both the femoral head and the hip socket. There are various names for the prostheses and the operation; we will call it the Total Hip Replacement or THR. Several British surgical centres were involved, and many designs and combinations of materials were tried with varying success. But by the later 1960s it seemed clear that the best model was that developed by John Charnley at Wrightington Hospital, near Wigan in industrial Lancashire, which had been a surgical centre for TB of the bones and joints.

By the 1970s hip replacement was a paradigm of surgical success, by the 1980s it was big business in the US, and by the 1990s the business was globalised. Which is not to say that hip replacement was or is always successful: the demands on this medical device are now high – not just that patients move with reduced pain, but that they can lead an active lifestyle.[8] The expansion of orthopaedic surgery has continued beyond the hip, and replacement surgery is now performed on the hand, shoulder, elbow, knee, ankle and spine. From its origins with innovative experimental surgeons working alone, the total joint replacement market is now worth about US$14 billion a year.[9]

The effects on the surgical professions have been huge. Most orthopaedists in the US and UK now specialise in the replacement of particular joints. Sports medicine grew out of orthopaedic problems, and is now a specialism in its own right, reaching beyond professional sportsmen to weekend joggers and occasional squash players. *The Journal of Arthroplasty* (which is largely devoted to joint replacement) was started in 1986, but hip surgeons were organised well before that, and published in the general orthopaedic journals such as the *Journal of Bone and Joint Surgery*. The Hip Society was started in the US in 1968 under the leadership of Frank E.

Stinchfield. Today it is a non-profit educational organisation that sponsors an annual forum for orthopaedic surgeons, and an open meeting for both surgeons and engineers at the Annual Congress of the American Academy of Orthopaedic Surgeons. The International Hip Society was formed at the same time.[10] The Association for Arthritic Hip and Knee Surgery, established in 1991, is another group of surgeons who concentrate specifically on hip-replacement surgery.

This book will explore the invention and development of the artificial hip. It will focus on the role of the 'professionals' – the surgeons and engineers who developed the prostheses and the associated equipment – but it will also include the manufacturing companies and regulation by governments. It will endeavour to map out this part of the 'medical device market' as it developed in Britain and the US, and it will explore the interactions of patients, professionals, companies and governments. It will ask how THR compares with other technologies within medicine, and how the developments have differed between western countries. Explaining such phenomena involves discussion of markets and industries, of patient and professional attitudes, and of modes of assessment and regulation.

We hope this book will appeal to all who are interested in the past and present of medicine or technology, and who want to understand medical innovations in historical and geographical perspectives. We would like this volume to be used by people with a professional or a personal interest in THR – for whom we have tried to include patients as well as doctors and engineers. And it should also help those who want to analyse the forces at work in the ongoing transformations of medical services, in the US and especially in the UK.

A brief review of some useful books and articles

Most of the research for this volume came from original (or primary) sources including the leading orthopaedic journals, or from discussions with surgeons and other medical professionals. We looked closely at the major British innovators, but especially at the career of John Charnley, who published hundreds of articles and four major texts on orthopaedics, two of which related particularly to hip-replacement surgery: *Acrylic Cement in Orthopaedic Surgery* (1970) and *Low Friction Arthroplasty of the Hip: Theory and Practice* (1979).[11] Both volumes provide detailed description of Charnley's techniques and their rationales.

Unlike some other well-known surgeons, Charnley did not record his own life-story, but William Waugh wrote a lucid, in-depth biography, focussing on his hip replacements. Our, complementary, study looks more widely at the British sites of innovation, at the development of the THR after Charnley, and especially at the interactions of surgeons, industry and government in the UK and the US. For the latter, one outstanding source is *Bare Bones* (2003) by Augusto Sarmiento, who recounts his career as an orthopaedic surgeon, focussing on the increasingly problematic relationships between device manufacturers and surgeons.[12]

Design and materials are an important part of the hip-replacement story. Though, generally speaking, there is little historical research on metals, plastics and ceramics, there are some useful sources on orthopaedic materials, including an edited collection on prosthetics, in which the article by Alex Faulkner highlights the pressures placed on engineers.[13] Dane A. Miller, who is President of Biomet, an international device manufacturer, used his engineering background to write a detailed history of orthopaedic products in Leslie Klenerman's book *The Evolution of Orthopaedic Surgery* (2002)[14] – a collection of historical essays expertly covering replacement technologies for the various joints.[15] The section on the development of the THR, written by Klenerman himself, offers a neat and insightful summary from the clinician's point of view.[16] Useful institutional histories include Charles V. Heck's, *Fifty Years of Progress* (1983), published in association with the AAOS.[17] A similar volume exists for the British Orthopaedic Association, also written by William Waugh to mark its 75th anniversary in 1993.[18]

Such are the kinds of sources available for historical studies of recent medical chronologies – original documents and interviews, biographies, celebration histories, technical histories and recent reviews by social scientists. We will come soon to the books and articles by historians that provided the immediate background to our own work, but first we want to draw attention to recent studies on medical technologies more generally. From about 1980, a growing band of researchers in the history of medicine and in various social sciences, in various countries but perhaps especially in the UK, have examined a range of technological innovations.

The industrial and professional history of the heart pacemaker and later, the defibrillator, is discussed in detail in Kirk Jeffrey's book, *Machines in Our Hearts* (2001).[19] In this book, Jeffrey showed how cardiac-pacing developed through competition and conflict between doctors, engineers, industrialists and 'outside groups' such as the Food and Drug Administration. He examined the market for these new products, the lawsuits over competing patents, the aggressive marketing to the medical profession, and the entry of company salesmen as technical consultants in laboratories and operating rooms. In a study of four types of medical imaging, Stuart Blume's book, *Insight and Industry* (1992), highlighted the choices made by doctors and firms as new technologies built on older ones, shaped by the industrial and professional patterns associated with previous methods.[20] His book has been formative for our work, and we use it especially in the last chapter. In *Technology in the Hospital* (1995), Joel Howell provided a wider picture of the uses of technology. He reflected on the ways that technologies were haltingly adopted in American hospitals in the early 20th century, how they fitted into the increasingly organised and standardised institutions between the wars, and how they came to characterise hospital care after WWII.[21] All these three volumes explore the nuances of medical technologies in specific locales and discuss the role of technology in medical marketplaces as these expanded beyond pharmaceuticals and direct doctor-patient relations.

In the decade from 1992, there were four collections of essays on the history of medical technologies, edited by John Pickstone, Ilana Lowy, Ghislaine Lawrence and Jennifer Stanton,[22] plus detailed reviews of the literature by Jennifer Stanton and Harry Marks.[23] In large measure, all were products of the growing interest in the social history of medicine, and of a shared determination that the technical aspects of medical sciences and technologies should not be regarded as 'beyond the social'. To be sure, the historian had to understand the technicalities of the innovations, but not just in the ways they were understood by scientists or clinicians looking back to the roots of present knowledge. The historians had also to understand that the knowledge and technologies might have changed their forms and meanings over time; and that the contexts in which they originated and developed were not necessarily the contexts of our present. For example, historians of medicine grew sensitive to the fact that medicine in war was different from that of peacetime, and that many of the professional gains in the Great War, say, were lost when the Red Cross hospitals were disbanded.[24] Or that there were different ways of understanding the effects of oxygen on the body, so that different groups of doctors might disagree about its use.[25] Or that the early use of cortisone for rheumatism was also an exploration of the meanings of a 'wonder drug' and an experiment in the management of hope and of disappointment.[26]

The volume edited by Ghislaine Lawrence especially reflected the increasing interest of historians in the displays and collections of *museums* – in what could be learned from medical artefacts that was not evident from paper descriptions and pictures; or how artefacts could be used to stimulate public interest in medical technologies. This link between museums and medical technology studies has continued to develop. Indeed, it has provided one happy outcome of our present research. One of the authors, Francis Neary, was jointly responsible for a temporary exhibition on THRs mounted in the newly renovated Humterian Museum at the Royal College of Surgeons in London, and for two displays at the Wrightington Hospital – one on Charnley and one for the education of potential patients.[27]

Oral history also features in our study: we have interviewed and held discussions with more than 70 surgeons, patients, engineers and manufacturers over the course of our research. Comparing and contrasting the testimonies of surviving 'actors' proved invaluable in reconstructing the early professional networks and discovering how research was carried out on a day-to-day basis at different centres. For instance, our interviews with Harry Craven (Charnley's first engineer), revealed how impressive results were achieved at Wrightington with improvised machines, often built from parts rescued from a local scrapyard.

Through other interviews and our attendance at orthopaedic conferences, we discovered some of the preoccupations of present-day surgeons, engineers and manufacturers and how these built on past debates and controversies. We learned that many of the concerns of the pioneers live on today, for example in debates over the size of the femoral head or whether to use cement; we also saw how opinions on such matters often depended on the place of training. The oral history research for this project led to a Wellcome Witness seminar on the early development of THR, where many of the surviving actors in the early story gathered at the Wellcome Trust in London in March 2006 to recount their experiences.[28]

But if we return now to the historians' accounts, we see that the edited collections from the 1990s also evidenced a convergence of the social history of medicine with the new history of technology, as established in the US from the 1970s. Through this interaction, historians of medicine were alerted to issues of gender in technologies, to new ways of relating technologies to company histories, and to the importance of technological 'systems', especially for urban infrastructures.[29] Of particular interest was the link with economic studies, especially the work of Nathan Rosenberg, including debates about the different conditions for innovation in the US and the UK.[30] The link with economics and economic history came to include patterns of selection, convergence, and possible 'closure' among the variants of new technologies, together with issues of path dependence.[31] From sociology of science came studies on the social construction of technology, and how artefacts could bear many meanings and uses, especially in their early stages.[32] All these perspectives have proved useful for later studies of medical innovations, including this one.

These approaches have also fed into the growing number of sociological researches sponsored by research councils concerned about the impact and acceptability of new biomedical techniques. Federal agencies in the US have supported many such studies.[33] A good British example is the programme of research on Innovative Health Care Technologies funded by the ESRC and led by Andrew Webster – a programme which supported most of the work for this book. A collection of essays from that programme has now appeared, and it includes a study from Manchester comparing the development of the THR with that of the replacement lens used in operations for cataracts.[34]

The most recent collection of historical essays on medical technologies has been edited from Manchester by Carsten Timmermann and Julie Anderson.[35] These studies are perhaps less concerned with questions of technological evolution and policy, and more with questions of locality and social relations. The volume contains an essay by Anderson on Charnley and the prevention of infection, which nicely shows how Victorian questions about the source of germs and the role of the air were renovated by post WWII hip surgery.[36] A companion essay by Anderson published in a journal collection on regional history, underlines the importance for Charnley of local, industrial and academic connections – again a theme we take up in this book.[37]

This link to locality and to industry brings us to the two studies which particularly helped create the context for our own researches. One was a project on orthopaedics in Britain from the later 19th century to WWII, which sprang from Pickstone's interest in the orthopaedics of Manchester and Liverpool but which broadened into a national study. Roger Cooter beautifully explored the conservative approaches to surgery which characterised the leaders of British orthopaedics in the first half of the 20th century. These creators of a new surgical specialism stressed immobilisation, physiotherapy and the restoration of function. Their ambitions were assisted by the special orthopaedic services formed in the Great War, which also involved leading surgeons from the US where surgical specialisation was more advanced.

Cooter traced the Liverpool tradition in orthopaedics from bonesetters who settled there from Anglesey, the island off the northwest of

Wales. He looked closely at Huw Owen Thomas, the politically radical son of a bone-setting family, who qualified in (orthodox) medicine and worked as a general practitioner by the Liverpool docks, attaining fame for his splints. His nephew, Robert Jones, became a consultant surgeon and specialised in orthopaedics. As a young man in the 1890s, Jones organised the casualty services for the digging of the canal which would take ships to a new port in Manchester, 35 miles inland from Liverpool. That experience served him well in the Great War when he led the military orthopaedic service, which included several leading American surgeons. It was Jones' protégé, Harry Platt, who developed the orthopaedic tradition in Manchester, first in the industrial district of Ancoats and then at the teaching hospital, the Manchester Royal Infirmary.[38] Sir Harry, who lived to be a hundred, was an inspiration and a ready source for the historical work in the 1980s by Roger Cooter, John Pickstone and Stella Butler.[39] He was also the teacher and patron of Charnley, and partly responsible for securing Charnley his research base at Wrightington, where he (Platt) had worked on TB surgery alongside the orthopaedists from Liverpool.[40]

The other key work was not on Britain, but was equally rooted in a deep understanding of a particular local and national context. Thomas Schlich was engaged at the medical history department of Freiburg University, to write a history of the AO (Arbeitsgemeinschaft für Osteosynthesefrage, or the Association for the Study of Osteosynthesis).[41] The AO were a group of Swiss surgeons, formed after WWII, who revolutionised fracture care through the insertion of metal plates screwed to the bone. Like partial joint replacements, this 'internal fixation' of fractures had a prewar history of failure. The postwar programme benefited from better materials, antibiotics and the other aids to surgery we mentioned above, but it became a generally available technique because this tightly organised group of Swiss surgeons carried out a systematic programme of research and development, directly comparable to the industrial research of large companies. In their case they franchised the production of the metal parts to a small industrial company which became part of their extended AO system – which also included training courses and text books to disseminate their practices.

Schlich analysed how this unique professional-industrial structure was related to the politics of Switzerland; he linked it with TB sanatoria and with skiing injuries, and showed how the technology developed with an apparent continuity, belying the fact that the initial biomedical principles were abandoned. He also explored how the AO fared in different nations and markets; some of their greatest successes were in communist states where the rapid uptake followed the conversion of the government to the benefits of the system. Initially, they fared less well in the US, for reasons we will analyse in relation to THRs.

There is much in Schlich's book which connects with this present study, not least the creation of a surgical system where all aspects of the prosthesis, the operation and the environment were carefully controlled. This corresponds well with Charnley's work, and indeed the chief mover of AO, Maurice Müller, was a friend of Charnley; he was also a major developer of Charnley's THR innovation. In fact, Müller's modification of Charnley's design succeeded better than the original in the US. Here, as so often, comparative history also become the history of connections, and of international networks which worked differently in various contexts – and which also helped to shape those contexts. That is the kind of approach we have taken here, first for British networks, and then for their international connections.

Chapter outline

Chapter 2: Surgeons and Engineers in Postwar Britain: Medical Technology in Local Contexts

Partly to put John Charnley's work in a wider context, and not to focus too exclusively on the 'winner' among the British innovators, our first substantive chapter, after sketching the technical background to THRs, then delineates the five major British programmes to develop hip implants, three of which preceded or ran alongside Charnley's work. We compare the aims, resources and outcomes of these programmes, and especially their use of technicians or engineers. We analyse the local determinants and explore how a range of designs was created, which was only partially reduced by convergence towards Charnley's design.

The pioneering British programme was led by Kenneth McKee at Norwich from the late 1940s. He had been inspired by the prewar work of Phillip Wiles at the Middlesex hospital in London.[42] In the Norfolk and Norwich Hospital, which was not a teaching hospital, McKee mostly worked alone, with some help from the City Engineer, and then with another surgeon who helped devise the McKee-Farrar prosthesis. The metal-head and metal-cup designs went through many modifications, but by the 1970s they were overtaken by the Charnley model which had fewer failures from breakages, loosening and infections. However, some McKee prostheses were found to survive very well, and metal on metal joints were revived in the late 1980s with more exact techniques for engineering the surfaces.

As we have already indicated, the most successful of the British programmes was developed at Wrightington, near Wigan, by John Charnley, who had trained at Manchester but moved out of the main teaching hospital in the late 1950s to set up a special research and development unit. His innovations were a new design for the femoral head and the use of plastic for the socket cup. Initially plastic proved disastrous; but a much better material was found and by the later 1960s the new design had become the world standard.

The Stanmore hip was developed from 1956 at the Royal National Orthopaedic Hospital, near London, within a novel department devoted to new medical technologies. The programme was much more sophisticated than Norwich or Wrightington in its systematic testing of materials and comparisons of existing techniques. The Royal National Orthopaedic Hospital had resources to compare and adapt, and the Stanmore design remains popular with very little variation from its initial form.

From the early 1960s at Redhill in Surrey, Peter Ring worked with a technician on his total hip design – like McKee he used metal on metal, but with fixation by screws. He refrained from using cement until the 1970s, and his concept of a cementless device became standard for younger patients from the 1980s. When he needed more resources than his small surgical company could provide, he was supported by Zimmer, one of the larger American device companies. His prosthesis attracted some continental users, but there were problems with his later use of plastic sockets, including their removal if the operation had to be revised. This programme, like McKee's, did not survive the retirement of its principal.

The final British programme examined here is the Exeter hip, which was developed by a surgeon and an academic engineer, with the support of the British branch of Howmedica, an American firm which had helped pioneer femoral head replacements. Unlike the other programmes, the Exeter research was begun after Charnley's success with plastic (acetabular) cups, and was essentially a modification of his designs, using a more popular surgical approach.

Overall one sees a shift from surgeons working with mechanics and engineers, to full collaborations between surgeons and specialist biomechanical engineers, with more active involvement of the companies. But we are interested in contrasts as well as trends and this chapter provides an opportunity for exploring the different ways in which the various groups defined and attacked the common problems.

Chapter 3: The Industrial Development of British Hip Replacements and the Charnley/Thackray Relationship

Here we open with a sketch of the (mostly British) industrial companies which were involved in the THRs discussed in the last chapter, but we then focus onto Charnley's work and especially his relationship with his manufacturing company.

Charnley worked closely with Chas F. Thackray Ltd of Leeds, which had supplied him with surgical instruments. For the THR, they forged the steel parts which he initially 'fitted' on his lathe at home. When his work attracted wide attention after 1965, he insisted that the special instruments and prostheses could only be sold to surgeons who came to train with him at Wrightington.

We explore how Charnley's designs were rapidly copied, especially in the US, though he and Thackray had tried to prevent this by not showing the prosthesis in the early catalogues.[43] The competitors usually charged much more, in part because British prices suggested low quality. The stainless steel that Thackray used was considered inferior to the more expensive cobalt chrome alloy favoured by American device manufacturers, and Charnley's adoption of polyethylene for the acetabular cup was also criticised. After 1969, Charnley and Thackray began to patent new features, and to produce variants to try to gain a stronger foothold in the American market.

Our aim here is to anatomise an extended programme of surgical research and development, and to place it firmly in a local context. We dissect Charnley's collaboration with a manufacturing company, revealing the tensions in a relationship still dominated by the surgeon. And we begin to see how the prostheses and the operation were imported to the US. The innovation succeeded there and Charnley was a hero to many American surgeons; but his own practices did not travel well, and Thackray's products hardly at all.

Chapter 4: Medical Technologies and Industrial Dynamics: The United States and the Global Market

Here, in a parallel with Chapter 3, we first discuss the pedigrees of some of the American manufacturers, many of them established in Warsaw, Indiana – a small town which became the world capital of the joint replacement industry. As well as disentangling the historical relationships between the various medical device manufacturers, the chapter also notes their associations with pharmaceutical giants. From the late 1980s there was a general consolidation of companies so that the US market, comprising some 160,000 operations per year, came to be dominated by Zimmer (26%), Stryker/Howmedica (20%), DePuy (22%), Biomet (10%), and Smith and Nephew (9%).[44]

From the 1970s, device companies competed through constant product innovation, especially around new materials; they also secured 'loyalty' from surgeons by various kinds of sponsorship and training, and with hospitals by exclusive deals. Company representatives became closely involved, even in the operating rooms. Though there is good secondary literature on pharmaceutical companies, especially in the US, and pioneer studies of particular medical devices and the companies which made them,[45] there is relatively little historical research on the industries which supply high-tech products to hospitals and clinics. So this chapter is important, not least because the medical companies based in the US are increasingly shaping the delivery of medical procedures throughout the world.

Chapter 5: Change and Stability in Technical Systems: Materials and Environments

Having discussed the industries, we return to questions about the evolution of the technologies. We detail the development from the 1960s to c. 2005 of both the main prosthesis components – the acetabular cup and the femoral stem – as well as the modes of fixation. Then we move out from the prosthesis to see how Charnley and others developed means of reducing the incidence of infections.

In 2004, the (new) National Joint Registry for England and Wales included 72 different brands of acetabular cup and 81 of femoral stem, available from 23 companies.[46] We ask why so many designs persisted, why there is little 'closure' onto one design, especially for the younger and more 'demanding' patients.

In some part, this lack of closure is due to differences in the patients. As the range of patients receiving THR expanded throughout the 1980s, there was some differentiation of designs according to the age, weight and activity level of the patient; and as more primary THRs wore out or loosened and required reoperation, there was further differentiation for 'revision' surgery. Some variation depended on the operative approaches preferred by different surgeons. But variety was not just inherited, it was sought. There was a long tradition of surgeons modifying existing instruments and naming the varieties – which may then be perpetuated at least among pupils and associates. The emergence of orthopaedic companies which developed designs with 'their' surgeons, added a new level of competitive commerce and lots of new variety – much of it barely tested.

1.3 Some of the Varieties of THR Femoral Prosthesis

Other persistent differences seem attributable to national contexts. In Britain, the NHS stressed economy and catered, at least initially, to patients with low mobility and low life expectancy. In the US, there was less concern with costs, and more with possible litigation, technical sophistication, and catering to younger patients. So the risks from the effects of cement, or of the wearing of metals and of plastics tended to be assessed differently in the two countries. The apparent risks also changed across time, as when new techniques of measurement allowed the better identification of wear particles.

In the section on infection control, our main concerns are the systematicity of Charnley's work and the way he resurrected old questions about the sources of germs. But also at issue here is his reluctance to rely on the prophylactic use of antibiotics, which became common in the US. Again, the relative costs and benefits changed over time. Charnley's concern with germ-free environments now seems both prescient and foundational as surgeons wrestle with the consequences of antibiotic resistance, even for operations much less vulnerable than hip replacements.

Looking closely at the many parameters of THRs, and their interactions over time, gives us a chance to analyse a system that does not converge onto one best solution – though some parts of it may do so. Such open systems are surely important, for the history of technology generally as well as for medicine. We tend to think of *sequences* of technical innovations; perhaps we should think more about models of competition, with slow feed-backs and many long-term uncertainties and interactions? And if the determinants are both technical and social, we might ask if a different medico-industrial structure would have produced different results?

Chapter 6: Hips, Health Services and Quality

We have already moved from the materials of the prosthesis to the environment of the operating theatre. Now we look at the changing place of THRs in the health services and health economies of the UK and the US, and then at issues of governmental regulation and registers. Throughout this chapter the national contrasts are strong, but so also are the transatlantic connections. If the early years of the THR were a story of British products taken up in the US, most of the stories in this chapter are of practices and ideologies moving eastward to Britain, and to the developed world at large.

In Britain, the history of THRs reflected that of the NHS under which they had been produced. One of the strengths of the new national service was its support for research, including surgical developments. Another was the free provision of devices such as spectacles and hearing aids, manufactured in national programmes. The devices were not elegant, but they brought assistance to many. They were part of the context in which Charnley devised a hip prosthesis which was cheap and reliable, at least for patients who were undemanding, both biologically and psychologically. And, as for most NHS procedures which proved effective, the problem then was the provision of sufficient specialist surgeons, and sufficient operating capacity to meet the demand. So hips became a prime example of long NHS waiting lists and the need to reduce them – as they remain to the present.

For the US, discussion tends to centre on the private market, though more than half of the THRs are paid for by public authorities, mostly by the Medicare programmes established in 1965 for the over 65s. Both for participants and for commentators, the key American issues were initially about the expansion of the THR market, about the roles of the companies, and about price inflation. Then came attempts to reduce expenditure, especially through fixing the Medicare payments for THR operations rather than paying 'reasonable costs'.

Most of the pioneers of THRs kept excellent records, but when many new devices were launched, and especially in the US, they were often used by surgeons without much prior experience, and the follow-up was minimal. Though there are bound to be tensions between innovation and regulation, these must be especially worrying where there is pressure to produce minor variants that are heavily marketed, using clinical reports which are statistically weak.[47] One answer is the use of local or national registers for the analysis of existing practices. Sweden had the first national hip register in 1979, but Britain waited until 2003, and the US has yet to follow, although individual programmes have been running many years.[48]

The push to trials for pharmaceuticals came in part from regulating agencies, especially the FDA in the US, but medical device industries were less regulated until 1976, in part to assist the development of a 'fledgling industry'. Much heavier regulatory measures have been enacted in the 1990s. In Britain, medical devices were not regulated in a meaningful way until the 1990s, although the British Standards Agency had regulated the quality of metals used for hip-replacement surgery. Closer regulation might have provided an early warning of some design problems, notably the Capital hip, which was sufficiently similar to established models as not to need extensive clinical trials in the UK. The subsequent enquiry by the Royal College of Surgeons concluded that 'there is no such thing as a small change in design'.[49]

Again, we have the chance to see how the professional assessment systems worked, how medical industries used systems and regulations that applied to most consumer-goods, and how the regulation of long-term devices compared with that for drugs. But we also look at the special characteristics of orthopaedic implants and how they came under national regulation as the contexts of judgement changed rapidly across nations, and as governments adopted new responsibilities for the quality and quantity of *care*, as well as for commodities.

Chapter 7: The Changing Nature of Patients: Expectations and Information

Issues surrounding markets and of consumerism bring us to consider the changing nature and expectations of patients, and how those related to the wider development of medicine over the last half-century.

Patient expectations have risen, and now relate more to 'normality' than to the preoperative condition. Few British patients have any idea what kind of prosthesis they are fitted with, though in the US direct-to-patient advertising on manufacturers' websites is becoming more common, so 'brand awareness' may increase. Unless the implant failed, patients forgot about their hip components – which were less 'experienced' than their spectacles or false teeth.

But the patients have changed, as well as their expectations. Charnley's first patients were mostly small elderly women who would not place a burden on a technology in its experimental phase. As further research was conducted into the type of stresses that an implant could bear, and new materials and designs were developed, the patient cohort widened to include much heavier and more active patients. It widened still further as younger patients also became the recipients for hip replacement. Thus cosmetic factors and quick rehabilitation entered the surgical desiderata, and hips were more obviously gendered.

This chapter explores how the body of patients, as it were, changed over time and between countries. We examine the ways that patients have gained information. We, like them, used a wide range of sources, from patient booklets provided by surgeons, to the internet and chat rooms.

Chapter 8: What is Past is Present

The book ends with a set of reflections on the ways in which THRs have changed lives, professions, industries, and health services; and on what we can learn from the history of THRs about wider issues in the evolution of medical systems, especially for the UK.

2
Surgeons and Engineers in Postwar Britain: Medical Technology in Local Contexts

Introduction

Before WWII, operations on arthritic joints were a last resort. Most involved cutting away part of the joint and fixing it (arthrodesis) so as to prevent motion and so reduce pain.[1] In some cases, the operated joint was manipulated to allow a certain amount of motion.[2] But the limitations of fixing joints were clear, especially to American patients, and it could not be done on both hips.[3]

> From the point of view of therapy, undoubtedly, arthrodesis as performed by us has given the best results, but, while this is a procedure which is readily accepted in Europe, in America we have difficulty in convincing our patients of the benefits to be derived from a stiff hip.[4]

In the 1920s the other main choice was excision. The diseased head of the femur was removed and the stump was moved up to create a moveable false joint.[5] But the surgery was not always successful, pain was not relieved and some patients were unable to walk afterwards. Attempts to replace parts of the joint with artificial materials almost always failed, and in operations that cut deep into bones, there was a high infection risk, especially when new parts were introduced.

Between the wars, the main business of orthopaedic surgeons, beyond injuries and congenital deformities, lay in the treatment of tuberculosis of the bones and joints. This could be operative, but conservative treatments were preferred – often involving long periods of immobilisation or elaborate exercises or dietary and light therapy.[6] Similarly, fracture treatment was conservative, using Plaster of Paris rather than attempting to secure the bones by inserting metal plates. Conservatism was then the modern form of orthopaedics, relying on physiology rather than mechanical insertions.

It was against this background that a few orthopaedic surgeons began to develop replacement parts – a project which became more popular after WWII, and especially in the 1950s, when it became clear that new antibiotics were helpful for surgery, and were also the basis of successful 'chemical' treatment for tuberculosis. Orthopaedic surgeons might be more adventurous, and their TB work was diminishing, freeing both time and facilities.

This chapter will review the British sites where surgeons and various kinds of engineers worked to create total hip replacements or THRs. We will ask about the particular aspirations and suppositions which lay behind the different projects, and how these affected the outcomes. The next chapters will say more about the manufacturing companies, in the UK and abroad; here we concentrate on how different THR designs were created and developed, and how they were related to the more or less contemporary development of single part replacements (or hemi-arthroplasties).

One-part and two-part hip replacements

The first THR in which both the femoral head and hip cup were replaced with metal prostheses was carried out by Philip Wiles (1899–1966) at the Middlesex Hospital, London in 1938. He operated on six patients with juvenile rheumatoid arthritis and one was still walking without pain 13 years later.[7] But the war interrupted this project, and more generally it diverted British and European surgeons to casualty work. After the war, Wiles' work was resumed by one of his trainees, and several other British surgeons began to try their own designs. By then it was possible to build on considerable, international experience of more limited interventions, either replacing the femoral head alone,[8] or interposing metal cups between the femoral head and the hip socket.[9] This work, from the 1940s and 1950s was mostly by American surgeons working with American companies.

In 1943, Austin Moore (1899–1963) of South Carolina and Harold Bohlman (1893–1979) of Baltimore replaced a cancerous femoral head by screwing side plates to the outside of the femur, the same fixation method as Wiles had used.[10] They later introduced a stem going down the bone canal to secure the metal prosthetic head, and this became the standard method. In September 1946, in Paris, the brothers Robert

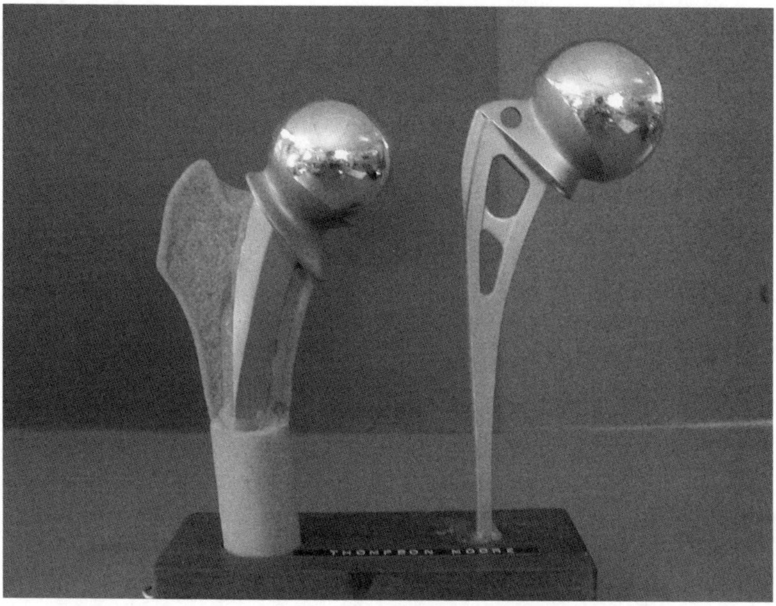

2.1 The Thompson (left) and Austin Moore (right) Half-replacements for the Femur only

(1901–80) and Jean (1905–95) Judet implanted their first mushroom-shaped acrylic prosthesis to replace the head of the femur.[11] Four years later they had implanted 600, but most of them failed as the plastic wore and broke – partly because the stem followed the axis of the neck of the femur rather than that of the bone canal. In 1950, in the US, Frederick Thompson (1907–83) of New York developed a femoral prosthesis with a stem made from chrome-cobalt alloy.[12] In 1952 Austin Moore modified his earlier design to create a self-locking, femoral replacement, the stem of which had holes to encourage bone in-growth.[13] Both the Thompson and Austin Moore femoral prostheses were successful solutions in patients with a fractured neck of the femur, and they were important prototypes for the early THRs.

Where the patient's hip socket was diseased, the standard remedy was the Smith-Peterson Vitallium mould arthroplasty, though the initial success rate was only 45%, and 62.5% thereafter.[14] Essentially, a cup was fitted over the head of the femur, which then moved against the cup rather than the diseased socket. The invention was first described by Marius Smith-Peterson of Boston in 1925, when he used a glass cup. He later tried steel and plastics, but when chrome-cobalt

2.2 An early Smith-Peterson Cup made from Glass

alloy was introduced into dentistry and orthopaedics in the mid-1930s, it soon became standard for Smith-Peterson cups; Vitallium was a trade name for the chrome-cobalt. Smith-Peterson cups were introduced to Britain by London orthopaedic surgeon Scotty Law after WWII.

There was one American experiment with a THR – at the Hospital for Joint Disease in New York, where Edward Haboush (1904–73) also introduced a fast-setting dental acrylic cement for fixing his Vitallium prosthesis.[15] But the results were poor, primarily because he had used the acrylic cement for glueing the prosthesis in place rather than as a grout for transmitting the weight of the body more evenly onto the femur.

Partly because of the pioneering work of Wiles, it was in Britain, under the newly-created NHS, that the possibilities of THRs attracted a succession of inventive surgeons. Once surgeons were on state salaries, they were not so dependent on their private patients, and some of them could devote more time to the development of new operations. Most of this chapter is concerned with these British ventures, which interacted in various ways, but which also differed between the sites, depending mainly on the characters of the leading surgeons and the resources that they could secure. If we give details of some of these programmes, it is to convey some impression of the trials and errors through which progress was made. We contrast the sites, partly to see the development over time, partly to see the results of different combinations of surgeons and engineers or technicians.

Norwich: The McKee-Farrar prosthesis and the fate of metal on metal designs

George Kenneth McKee (1906–91) had trained under Philip Wiles and became his senior registrar. He made models of hip prostheses in 1940 but he was worried about their fixation, so he did not experiment with them in patients, and the War halted his research. But in the 1950s, on becoming a consultant at the Norfolk and Norwich Hospital, he built on Wiles' early work. His hospital was not attached to a university and it was not in general a place of research, but McKee was convinced of the potential of the THR and he made it his life's work. For many cases of arthritis, replacing the femoral head seemed to him only half a solution, especially where the hip socket was also diseased.[16] McKee championed the THR when others (including John Charnley) were sceptical of its feasibility and looked to better single-part replacements or excision techniques.[17]

McKee's faith and persistence was partly religious. In a revealing article for his local church gazette, *The Mancroft Review,* he expressed his view of the body and of natural laws:

> There must, to my mind, have been an original inventor of such a wonderful structure and I can see no apparent reason why the laws cannot be altered at the will of their originator.[18]

Disease was the product of evil and difficult to square with a wholly benevolent God, but 'we can fight against it by every means in our power, by every method known to medical science, and also by prayer'.[19] Hence his faith that his surgical innovations would work despite their relative novelty and the risks that they created.

McKee's was a personal quest, more than team research or commercial development, though he sometimes worked with his orthopaedic colleague, Watson-Farrar, and he collaborated with a number of engineers over the course of his research. For example, he enlisted D.H. McCraken from Norwich City Council Engineering Department to perform metal tests, and his second hip joint, made entirely from a chrome-cobalt alloy (Vinertia), was developed with the help of John

Bunyan.[20] The later Vinertia components were made by Deloro Stellite, a British firm which specialised in this alloy.

McKee's early operations were small in number and performed only on patients with serious hip deformities, disease or trauma. In 1951, he reported his THRs on three patients, each with one hip which had been fixed in place surgically and with serious disease in the other.[21] He described his 'ball and socket joint' to a meeting of the South-East Metropolitan Regional Orthopaedic Club. The socket was fixed into the hip by a central screw with four other screws radiating out; the ball was on the end of a rod which was inserted into the shaft of the femur and fixed by a plate on the outside with three screws. The prosthesis was very much of a novelty and, when he demonstrated it, the 'meeting was impressed by the ingenuity and engineering skill shown in this device, but some doubt was felt as to how kindly the bone would take to such a mass of metal'.[22]

This was the first of five stages that his prosthesis design underwent between 1951 and 1972. The prosthesis was made of stainless steel, and in two of these very early THRs the screws were also made of stainless steel and soon became loose. However, the third was fixed with chrome-cobalt alloy screws and lasted for three years in a 70-year old woman. It was this case that convinced McKee that total hip replacement was a practical possibility.

In 1953, McKee visited the US and was impressed by the stem design of Thompson's femoral head replacement.[23] He adapted his own cementless metal cup to fit a smaller version of Thompson's prosthesis and between 1956 and 1960 he inserted the system into 26 patients. The new cup was fixed by a large screw that was incorporated into the design; it was the first THR cup to be made from the chrome-cobalt alloy Vinertia. In this second stage there was a 57% success rate after a seven-year follow-up of 26 cases.[24]

When, in 1960, it became apparent that the loosening of the components was the principal cause of failure, McKee started using a small amount of acrylic cement to improve fixation and spread the load over a large area of the bone. It was McKee's colleague Watson-Farrar who suggested the use of the larger, standard Thompson prosthesis for the femoral component to decrease the risk of dislocation and be more effective in load bearing. Other changes followed, both in the femoral design and the method of fixation. The resultant third series was the first to be inserted in a significant number of patients, with a success rate of 58% after a eight-year follow-up.[25] Out of 100 THRs, the main problems were incorrect positioning of the cup and faults with

2.3 Kenneth McKee with his Mckee-Farrar Prosthesis

insertion of the femoral component; sepsis also played a role; some patients had fallen, and some were 'over active'.

The THR had moved beyond the experimental stage, and in 1965 McKee and Watson-Farrar reviewed their materials and designs. They maintained their faith in chrome-cobalt alloy as an extremely inert material that did not self-weld *in vivo*. They further adapted the Thompson femoral component so that it had a narrower neck and a perfectly spherical head; and over time a variety of cup and femoral head sizes were introduced. The success rates increased from 75% to 86%, and then to 90% in the operations done between 1968 and 1969.[26] By the early 1970s, after a further series of further design and operative changes, McKee was claiming good or excellent outcomes in 95% of his cases.[27] He thought that it was 'probable that the current models of the metal on metal type will give better long-term results than the metal on plastic models'.[28] But this optimism was short-lived. As we shall see, McKee's model was being overtaken by other designs, notably John Charnley's small metal head with a plastic socket. In 1972 McKee stopped inserting metal on metal hips, though some surgeons continued to insert these prostheses until the early 1990s.[29]

From the mid-1960s, the McKee-Farrar THR had begun to be used widely in Europe and North America. It was also modified to suit the needs and interests of other specialist hip centres, for example, in Paris.[30] Early results were promising: in Finland 55 out of 81 patients operated on were totally pain free[31] and a longer follow-up yielded only a few further complications.[32] A Swiss study focussed on the remarkable pain relief achieved by the operation but the mobility results were not as good;[33] it recommended that the operation should only be performed on patients over 65 and that in some cases the Russian metal on metal competitor, the Sivash prosthesis,[34] was preferable. In Glasgow, Peter Freeman achieved pain-free and stable joints in the vast majority of his rheumatoid arthritis patients.[35]

But in a second wave of results, more complications began to appear. In Canada a 1972 study across five general hospitals, including 403 arthroplasties by 23 different surgeons, showed excellent results in relieving pain and increasing the range of motion, and good results in increasing the ability to walk, but there were many early complications of loosening, infection and thrombo-embolic disease.[36] The study concluded that THR should not be done by the 'occasional hip surgeon' who was ill-equipped to deal with these complications. A study of 100 THRs in New York came to similar conclusions about the improvement to the patient's quality of life, but also found a high incidence of infection and loosening. It recommended that the operation should only be done on the elderly, infirm or patients with severe clinical problems which could not be solved in any other way.[37]

By the mid-1960s, as we shall see, the Charnley hip was being widely used and surgeons in the Orthopaedic Department at Oxford, UK, made an in-depth comparison. They found the Charnley marginally superior in terms of pain relief, function and range of movement; but from over 100 operations with each prosthesis, the McKee-Farrar yielded six complications of infection and loosening, while the Charnley yielded none.[38] To attempt to stem the tide of negative results and to provide instructions on how to revise failures, a paper was published from Norwich on the management of complications using the McKee-Farrar technique.[39] But by the mid-1970s many centres had abandoned the McKee-Farrar in favour of the Charnley, which had displayed better short-term results.

Medium-term follow-up studies on the original metal on metal McKee implants in Sweden (after seven years)[40] and Norwich (14 years)[41] showed that good or excellent results dropped to 50%. Much of the comparative discussion then focussed on the difference between metal on metal designs and the metal on plastic combinations which Charnley used,[42] but a long-term follow-up study over 21 to 26 years showed that in fact, the patient selection and technical factors surrounding the positioning of the components were more important in determining the long-term results of the McKee-Farrar THR.[43] Unlike Charnley, McKee had not restricted access to his prostheses or surgical instruments, and by 1973 the manufacturers of his Vinertia prostheses, Deloro Stellite Surgical, had produced a manual of his surgical technique.[44] That the McKee-Farrar method was taken up by many inexperienced surgeons may have been a major cause of the complications. Comparing one kind of hip with another was not easy, and it was even more difficult to assess the relative skills of surgeons. Charnley, as we shall see, was well aware of that general problem.

Worries about carcinogenic and toxic effect of *metal* particles also persuaded some surgeons to abandon McKee's metal on metal articulation in favour of the Charnley;[45] but by the end of the 1970s, some surgeons were becoming more worried about the long-term effects of *plastic* particles. In the 1990s, a 20-year follow-up comparison of the McKee-Farrar and the Charnley THRs in Sweden concluded that:

> The long-term results of the McKee-Farrar prosthesis are comparable with those of the [Charnley] low friction replacement in this series.

> Wear of the polyethylene bearing and accumulation of polyethylene particles in the periprosthetic tissues may become an increasing problem of metal on plastic replacements in the long perspective. Therefore, the second generation all-metal implants seem to be worth considering in patients with long life expectancy.[46]

As we shall see in Chapter 5, metal on metal hips were indeed staging a comeback.

Wrightington (Wigan): The Charnley Prosthesis, plastics and cement

John Charnley (1911–1982) came from a pharmacist's family in Bury, an industrial town north of Manchester.[47] He qualified at Manchester Medical School and became especially interested in orthopaedics during his military service in WWII. Back in Manchester, in the orthopaedic department headed by Sir Harry Platt (1886–1986),[48] he worked his way to a consultant's post, and he undertook extensive research especially on the effects of compression in causing bones to grow together. This work on the external fixation of fractures led to the development of a metal frame for the compression fixation of knees; and Charnley also developed operations to fix hip joints. Some of the work was on animals, but mostly it was clinical, albeit with very substantial analytical contributions from histology and from biomechanics. Charnley worked with the professor of osteopathology and learned the histological skills.[49] As an engineer, he was self-taught, with his own lathe, though he later collaborated with a series of academic engineers and at Wrightington he was ably supported by a succession of technicians.

Though initially sceptical about the possibilities of artificial joints, Charnley became interested in problems of joint lubrication. That Judet hips squeaked seemed to him evidence of excessive friction, which he thought he could lessen by reducing the size of the femoral head and by using a low friction plastic, Teflon, for the hip cup. At one stage he almost arranged to devote about half his week to systematic research, but partly for private reasons, he instead set up a clinical programme to develop artificial hips, revising the metal prostheses which were already sometimes used to replace the head of the femur, and exploring the use of plastics for the pelvic component. To this end, he set up a department in Wrightington Hospital near Wigan (between Manchester and Liverpool).[50]

The hospital had been developed by Lancashire County Council as a centre for treating TB of the bones and joints – which by the mid-

1950s were being successfully treated by antibiotics. Across the industrialised world, TB hospitals, and orthopaedic surgeons, were then available for other functions; the AO orthopaedic foundation in Davos, Switzerland, was based in a former sanatorium.[51] More research money was available after WWII than before, and in Britain, where the NHS had made hospital medicine into a salaried profession, a young clinician could make more time for research.

That Charnley was able to take over part of a former county council hospital was also a consequence of the nationalisation of hospitals under the NHS. His mentor, Sir Harry Platt, sat on the Manchester Regional Hospital Board and had considerable influence over the development of orthopaedics around Manchester. That Charnley received funding from 'Action Research' was probably due to Platt's influence in that charity. And that Charnley did not want to remain at the teaching hospital was in part also due to Platt – to his reluctance to relinquish control of his department, where his 'legacy' was divided between Charnley and another protégé, David Lloyd Griffiths.[52]

Charnley was convinced that the key to better hip replacements was low friction – to be attained by using small femoral heads and plastic for the sockets. He was puzzled by the early success of the McKee and other all-metal prostheses, and by the discrepancies between results *in vitro* and *in vivo*. He continued to find it 'quite baffling that these implants give excellent results in a very considerable number of instances despite the fact that with this design it is impossible to simulate satisfactory function in the laboratory'.[53]

At Wrightington, he set up a clinical unit and a workshop and appointed Harry Craven to give technical assistance. Craven was not a graduate, unlike many of the engineers who later worked with surgeons. He was apprentice-trained, a practical man who worked with ingenuity and limited funds and did not get written credit for his work. The 'engineer's workshop' at Wrightington was a room with a bench, a couple of vices, an old-fashioned drilling machine and a blacksmith's forge.[54] While Charnley was still working at Manchester Royal Infirmary, Craven was going to Charnley's house, in the attic of which they made hip sockets from Teflon plastic, two days a week. Craven would make the cups, soak them in sterilizing fluid overnight and the next day Charnley would put them into patients.[55] The lathe was very basic, but Charnley expected miracles from it; it took Craven 45 minutes to make a single socket. At the same time Craven was developing a spherical grinder (based on machinery developed for optical lens grinding) to make the femoral head a perfect sphere.

2.4 The Biomechanics Workshop at Wrightington in the 1960s

Craven's life was made easier in 1961 when a new workshop was built and equipped with a grant of £6,500 from the Manchester Regional Hospital Board.[56] Charnley then managed to free himself from responsibilities at Manchester Royal Infirmary, by allegedly swearing at a patient. He resigned at the ensuing enquiry, and this allowed him to devote all his time to his unit at Wrightington. It also put him in the enviable position of having almost total control over the research and development there.

Before Craven arrived, Charnley had experimented with an uncemented Teflon double cup covering the head of the femur and the surface of the acetabulum. The idea was motivated by the low friction properties of Teflon, but the plastic wore and tore. Between 1956 and

1960, he switched to a stainless steel femoral replacement and various designs of Teflon cup. He inserted 300 of these prostheses; initially they worked very well, but it soon became clear that there were problems with wear. Before the full extent of the difficulties were realised, Craven was asked to make a test rig to measure the effects.[57] On the rig, a Teflon cup wore out completely within 24 hours; and X-ray evidence confirmed that wear was taking place in patients. Within two years, 95% of the series had to be revised. Charnley was deeply depressed, but the patients were keen to have further operations. These early failures did not appear in print.

To combat wear, Charnley tried to further lessen the friction by reducing the diameter of the femoral head from 1 inch to 7/8ths of an inch, but with little effect. At this time there were three lengths of neck, but Charnley always insisted on using the short neck. Craven made a Perspex model to convince him that as the femoral head was wearing through the Teflon cup, the flange of the short neck was catching the periphery of the socket. The pelvic cup was so damaged in some of these cases that the only option was a Girdlestone's operation[58] to remove the joint – an option which, of course, the patient might have faced initially.

In 1959 cement was introduced for the femoral component, and later the cup was also cemented. Charnley had learned of the cement from Dennis Smith, a material scientist in the Dental School of Manchester University.[59] It worked well, but the wear problems with Teflon were not eased by this development. Craven designed a device with bellows that could be incorporated into the hip to supply lubricant, but they could not find a lubricant that was inert in the body. Charnley was reluctant to discontinue the use of Teflon, which at that time had the lowest coefficient of friction known; it seemed a move away from the first principles of his 'low friction arthroplasty'. However, he was forced to look for a more resilient material from which to fashion the cups.

Between 1960 and 1961 he experimented with Fluorosint, a very hard plastic that had just been developed. In exchange for samples, Charnley swore not to divulge its formula. Craven was sceptical about its utility, for when he came to machine it, the plastic wore the steel tool. After two days on the test rig, lubricated with distilled water, it had worn a groove in the stainless steel. In experimental patients, it proved a good bearing surface, but it wore itself and the stainless steel head, leading to a considerable build up of particles and eventual loosening.[60]

The Christmas of 1960 was a difficult time. Charnley's experiments with alternative plastics had failed and he thought deeply about putting patients through further experiments, with the possibility of still not developing a successful operation. He seriously contemplated abandoning his work on the hip.[61]

Early in 1961 Charnley went to Zurich to visit his Swiss orthopaedic colleague and friend Maurice Müller – to gain inspiration and give a series of lectures. Just before he left, a salesman from Todmorden, another industrial town, called on Craven offering a new plastic for sale. Craven was ready to send him away but decided there would be no harm in asking for a sample, which he put on the test rig. When Charnley came into the workshop and asked what it was, Craven explained that it was Ultra High Molecular Weight Polyethelene, otherwise known as high density polyethylene or HDPE. Charnley got into one of his tempers and told him to throw it away; he was through with plastics and expected solace from Müller's metal on metal implants in Zurich.[62]

But while Charnley was away, Craven carried on testing the material both on the test rig, and on another device he had developed using the rocker shaft of his car. He knew from his tests on Teflon, that 24 hours on the test rig was approximately equal to two years in the body. But after two and a half days on the rig, the HDPE had worn by only half a thousandth of an inch – far better than any other material available; and it did not wear the stainless steel head at all. On his return, Charnley still needed convincing, especially as the frictional properties of HDPE were much poorer than Teflon; but Craven's wear graphs and the manufacturer's specification gave him no option but to try it in surgery.[63] The results of the first operations were excellent.

Charnley later developed a pendulum test rig to show visitors how his metal on plastic articulation was far superior to the metal on metal articulations developed elsewhere. Many visiting surgeons were convinced when the metal on metal pendulum came to a grinding halt, while the metal on plastic one carried on swinging indefinitely. In fact, the lack of lubrication rigged the results in favour of the metal on plastic joint, and the experiment was propaganda for Charnley's own hip rather than a rigorous test of the coefficient of friction.[64]

All the components of the Low Friction Arthroplasty were now in place: the stainless steel femoral prosthesis with small diameter head, the HDPE socket, and the acrylic cement to fix them. The development was the result of a close collaboration between Charnley, Craven and others, plus lucky contingencies. The basic design remained unchanged

2.5 John Charnley with his Low Friction Arthroplasty Prosthesis, early 1960s

until the 1980s but a greater range of femoral replacement lengths, stronger stainless steels and a Vaquasheen finish were developed in the 1970s. As we shall explore in the next chapter, Charnley chose to work with a small manufacturing company in Leeds, Yorkshire, which had made instruments for him. Chas F. Thackray and Sons worked to his direction, and he was able to control the supply of the prostheses.

Although the operation settled into a routine and the patients with Teflon hips could now be successfully revised, Charnley still worried about encountering further failures. He delayed publishing his results with HDPE for as long as he possibly could. In 1968, after five years clinical experience and with some trepidation, Charnley was ready to announce his operation to the world.[65] But he worried about releasing the prostheses before the full five-year results were available, so he restricted the sale of the implants. The surgical instruments could be bought freely because they could be used with other operations, but before Thackray's would sell the prosthetic implants, Charnley had to approve the surgeon concerned.[66] They had to visit Wrightington for at least two weeks and show that they could use his favoured surgical procedure. This involved approaching the hip from the side and removing the greater trochanter, a ridge of the femur (with its muscles attached), so as to create a larger exposure of the hip joint for accurate placement of the prosthesis. Once the prosthesis was inserted, the greater trochanter was then wired back into place.[67]

Charnley was concerned that surgeons would use his prosthesis with other surgical approaches, and that if dislocation occurred, the surgeon would blame the prosthesis and tarnish its reputation. These first, idiosyncratic Wrightington 'courses' in THR established Charnley as a world authority, though by the late 1980s other centres like Exeter (using the posterior approach) were to steal the lead on Wrightington in developing more professional training-weeks.

Charnley also protected the results of his operation in another way: he refused to operate on patients below the age of 65, feeling that his prosthesis would not fare well in younger, more active patients, and he set about developing strategies for delaying the operation in younger people.[68] These precautions helped make the Charnley prosthesis the best performing model over five to ten years – after which a small number of HDPE cups began to wear out and loosen, due to particulate debris.

Charnley appears to have been uninterested in making money from his prostheses. He did not patent the original features of the ball and socket joint in the 1960s and his tools and prostheses were quickly copied in the US and Europe. We shall follow these later developments in the next chapter, but first we continue with the comparison of British sites.

The Stanmore prosthesis and engineering sophistication

The development of the Stanmore hip was the result of a collaboration between J.T. Scales (1920–2004) and J.N. Wilson (1919–2006). Scales was an inventive surgeon who set up the Department of Biomedical Engineering at the Royal National Orthopaedic Hospital at Stanmore and London; Wilson was a surgeon at the same hospital.[69] The Stanmore programme was initially characterised by relatively elaborate designs and specialist equipment, including the first hip simulator – rather more sophisticated than the Wrightington test rigs, and a precursor to the elaborate machines now used by major companies and their university collaborators.

Work began in 1956 on a metal on metal THR, just before Charnley started his programme. They devised a 'double' cup for the hip socket, and the best method of fixing the outer cup seemed to be three chisel-pointed nails.[70] To insert the nails a specially designed pneumatic hammer that withstood autoclaving was developed by the General Electric Company. To prevent the nails 'backing out' and interfering with the movement of the artificial joint, a lining cup was locked into the outer cup after fitting. A special instrument was designed for the removal of the inner cup after it had been locked in place. Scales and Wilson thought that nails would ensure more permanent fixation than cement because experiments on animals had shown that acrylic resins were subject to degradation and appeared in the urine about 44 weeks after an operation.[71] Their inner cup had a concave, horseshoe-bearing surface to facilitate the 'ingress of fluid to the inner bearing edges' and to provide 'an exit for any particles produced by wear'.[72]

The sophistication also extended to the femoral component. Most femoral head replacements, whether used alone or in THRs, could be fitted in either hip; but one sophisticated version of the Thompson involved separate left-hip and right-hip designs. The Stanmore followed this precedent, and the femoral prosthesis was fixed in place with a Simplex 'C' polymer that had a longer curing time (10–15 minutes) than the widely used Simplex 'P'. This allowed the cement to be poured into the marrow cavity (which was reamed out for about five inches), and it reduced the risk of infection caused by handling the polymer. Scales designed equipment to machine the bearing surfaces of both components and found that a ground surface was preferable to a polished. Fifteen THRs (mostly for osteoarthritis) were done between April 1963 and January 1966; none had to be removed or became infected.[73]

Further research at Stanmore concentrated on control of wear because of its profound effect on the efficacy and longevity of the device. At this time, when Charnley had just chanced upon HDPE, the Stanmore team (like most other surgeons) felt that plastic materials had severe limitations, given the high pressures involved in hip joints. They built their hip joint simulator and a pin and disc machine to test various metals – for friction, wear resistance and type of debris produced, under dry conditions and with saline or blood plasma. Titanium gave the worst performance in both tests and suffered almost immediate catastrophic failure on the hip simulator. Stainless steel did little better, but the chrome-cobalt showed minimal wear even when the load of the machine was increased to its maximum and the hip simulator had run 300,000 cycles with blood plasma as a lubricant. This material was chosen as the material for both the cup and femoral components, which were patented in 1966.[74]

In 1970 Wilson and Scales published a major study of the loosening of THR with cement fixation.[75] They indicated that patient pain was the best diagnosis of loosening and that loosening of femoral replacement was nearly always permanent, needing revision surgery. Loosening of the pelvic cup could be less serious and temporary.[76] But after recording the loosening of eight stems in a series of 67 Stanmore implants, and 13 loose stems out of 250 McKee-Farrar implants, they considerably revised their initial faith in the metal on metal prosthesis. They did tests on the Stanmore, McKee and Charnley THRs and found that the 'high loadings acting through the hip joint may in some metal-on-metal prostheses result in the development of frictional torque values of such magnitude as to disrupt the bond at the bone-acrylic interface either in the acetabulum or the femur.'[77] This was less likely with the Stanmore cup, secured with nails, but the values for frictional torque using polyethylene cups were considerably less than all the metal equivalents. Because of the seriousness of loosening, they came to question whether metal on metal replacements should continue to be used.

In the meantime, they had developed their own metal on polyethylene THR, which fared as well as Charnley's version in terms of frictional torque. They moved cautiously towards their new plastic, RCH 1000: 'If the future shows that there is no deterioration of

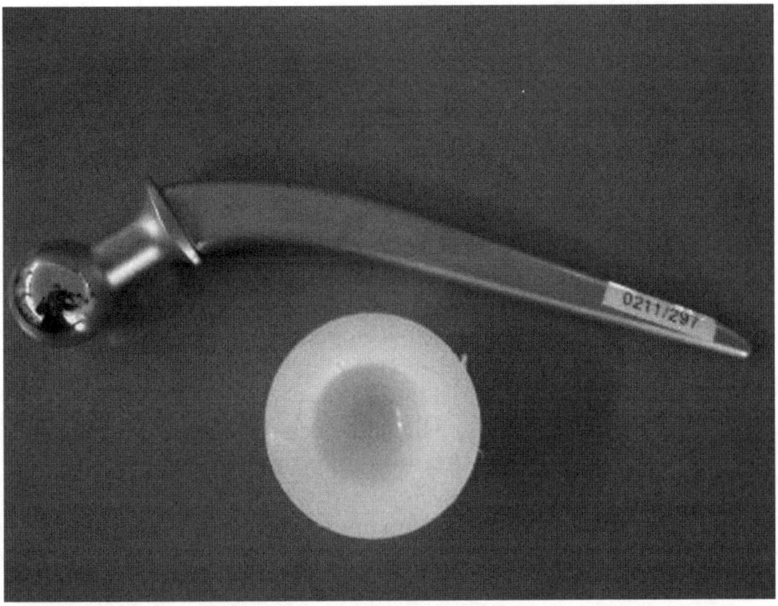

2.6 A Metal on Plastic version of the Stanmore Femoral Prosthesis and Acetabular Cup

the mechanical properties of RCH 1000 over the years, and that the wear products of the polymer do not produce an undesirable tissue response, neither adjacent or in the tissues remote from the implant, then metal-on-metal bearings will be discarded.'[78]

In 1980 the survivorship rates were analysed for 173 metal on metal prostheses, implanted between 1963 and 1972, and 248 metal on plastic implanted between 1969 and 1972. Failure was mainly due to loosening and the annual rate of loosening increased as the follow-up time increased, suggesting that it was wear-related. For the metal on metal group the overall probability of survival was 53% after 11 years and average annual probability of removal was 5.5%, compared with 88% after eight years and 1.5% for the metal on plastic. The change from metal on metal to metal on plastic was vindicated.[79]

The Stanmore has remained a popular model in Britain – a combination of original engineering and the adoption of polyethylene after comparisons with Charnley's models. In the teamwork of surgeon and specialist engineer, Stanmore helped develop a new approach to biomedical engineering in Britain. They worked with Zimmer UK, who arranged for the chrome-cobalt prosthesis to be manufactured by

Deloro Stellite, the specialist metals company which they later bought.[80]

Redhill (Surrey): The Ring prosthesis, a pioneering cementless device

In the 1950s, Peter Ring was the Evans Lowry Professor at the Royal College of Surgeons, where he performed pioneering work on internal fixation of fractures. In the early 1960s he moved to Redhill in Surrey to found a fracture clinic and the first orthopaedic and trauma service in the area. At the same time he developed his famous metal on metal uncemented THR. Like McKee in Norwich, Ring took London experience into a minor hospital which did not have the resources of the Stanmore hospital, nor those which had been created by Charnley out in Wrightington. Ring, too became known internationally, but chiefly as a pioneer of elaborate technologies which did not quite work, even when backed by the resources of American companies.

Rings' first model was similar to the McKee-Farrar prosthesis in that both the femoral and acetabular components were made of metal and not fixed by cement. The femoral component was made from chrome-cobalt alloy and based on the Austin Moore replacement for the femur only.[81] The main innovation was in the metal acetabular cup, which

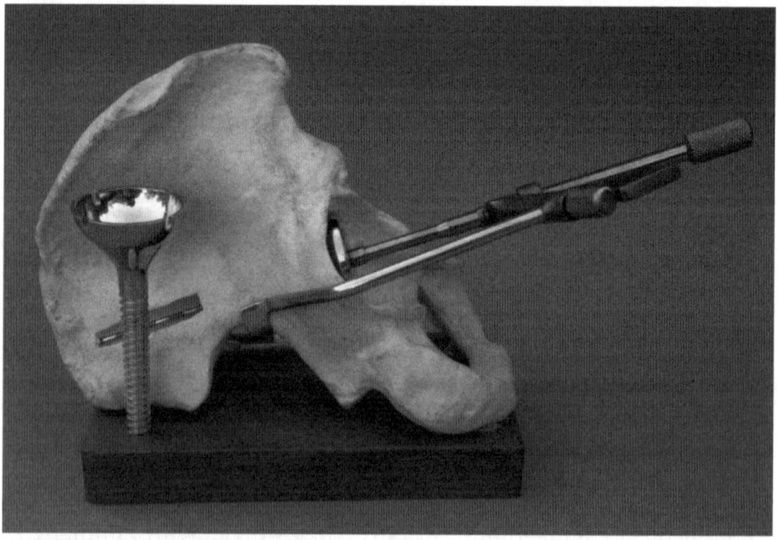

2.7 The Acetabular Cup and Jig (guide) designed by Ring

was fixed into the patient with a long-threaded screw. Ring's operative assistant, Peter Shepard, developed a highly specialised jig that ensured correct positioning of the long screw that went deep into the pelvis. Shepard had no formal engineering background. Like Craven at Wrightington he was a gifted mechanic, but with extensive experience of orthopaedic operating practice.[82]

In the late 1960s, Ring responded to the success of Charnley's operation at Wrightington by trying a polyethylene head on the femoral replacement. His aim was to lower the friction of the bearing surfaces, but the use of plastic on a convex surface proved disastrous and he quickly returned to the original all-metal femoral component. Under the guidance of Michael Freeman[83] (who saw no future in metal on metal replacements after the publication of Charnley's results), Ring switched in 1979 to a polyethylene cup fixed by means of an offset peg. It was meant to reduce friction and also ease the difficulty of revision surgery: the original metal cup was screwed so far into the pelvis and fixed so securely that it was extremely difficult to remove.

In the mid-1970s, Ring addressed the problem that his femoral component was not a good fit in the femur, and the square stem of his Austin Moore type design was cutting into the bone. Instead of moving to use cement to fix the femoral component, he maintained his faith in cementless fixation and changed the construction material. As the Cold War was coming to an end, the biometals market was flooded with titanium. It had the advantages of being lighter than stainless steel or chrome-cobalt; and bone would also adhere to its surface so its potential for cementless fixation was good. His new design incorporated a wire mesh in the top part of the stem of the prosthesis into which bone would readily grow. The development of the manufacturing process of the mesh and its welding to the main part of the femoral prosthesis required a large financial investment. His British manufacturing company, Down Surgical, was too small to make this commitment so after extensive negotiations Ring switched manufacturers, to Zimmer UK.

The first application of titanium for hip surgery in UK was marred by a series of complications. Titanium was not a very good bearing surface, so the head of the femoral component was replaced with chrome-cobalt. A modular system was developed in the early 1980s with a series of chrome-cobalt heads fitted to titanium stems.[84] Ring also developed a titanium metal-backed cup, but like his original long-screw chrome-cobalt cup, it was fixed so well (this time by the bone growing through the wire mesh) that it was very difficult to remove in revision surgery. To make matters worse, the titanium prosthesis had a tendency to fracture at the neck. After Ring retired in 1988, surgeons at Redhill were forced to revise many of his titanium and earlier implants and subsequently switched to the Charnley modular system.[85]

Ring had excellent or good results with 74% of 169 patients after five to eight years follow-up with his early models, and he improved this to 90% at one to five years with 535 patients who had the newer models used in the early 1970s.[86] By 1978 he had 14 years of follow-up results: the early models were showing 66% success rate and the later models 94%.[87] After 21 years these figures had fallen off to 59% and 77% respectively,[88] which was still impressive.

Ring's total hip system was not used as widely as the McKee-Farrar and Charnley varieties but had some supporters, especially in northern Europe (where there were concerns about particle debris from the cement and polyethylene cups leading to loosening and sepsis). The early follow-up results from other centres were mixed. Longer-term follow-up in Birmingham found 75 out of 154 hips had good or excellent results after 5–12 years,[89] and in Belfast a study of 250 Ring Mark II prostheses implanted between 1968 and 1974 showed that the cumulative survival rate was 61.4% after 21 years; the Ring here performed as well as the McKee-Farrar prosthesis and better than the Stanmore.[90] A Danish nine-year follow-up study reached similar conclusions, finding the overall probability of survival comparable to the survival rate of some commonly used cemented prostheses.[91]

Ring remains known as a pioneer of cementless prosthesis and for metal on metal hips, a path followed by several American surgeons and manufacturers as we shall see; but his own results failed to surpass the cemented models and revisions proved more difficult. Titanium, however, did prove popular, especially with the US company Biomet.

Exeter: The Ling-Lee hip, building on Charnley

The Exeter hip was the result of another close collaboration between a surgeon and an engineer, beginning in the late 1960s. Unlike the other UK models, it was deliberately intended to build on Charnley's work, including features for surgeons who did not share all Charnley's enthusiasms. The Exeter team worked initially with the London Splint Company, which was taken over by Howmedica (US) in 1972 after a long relationship between the companies.

After gaining his undergraduate degree in engineering from the University of Nottingham, Clive Lee worked for Rolls Royce on a jet engine reheating project in the early 1960s.[92] He subsequently returned to Nottingham to work on a doctoral thesis on experimental stress analysis, which was awarded in 1966. Later that year he joined the academic staff of the University of Exeter and began a fruitful programme of laboratory research with Robin Ling and other surgeons based at the Princess Elizabeth Hospital in Exeter. Ling had studied medicine at Oxford and after doing his initial training at St Mary's Hospital London, he specialised in orthopaedics at the Royal National Orthopaedic Hospital (Stanmore) from 1955–56. After some time in Canada as a Rockefeller Scholar, he completed his training in orthopaedics at the Princess Margaret Orthopaedic Hospital and Royal Infirmary of Edinburgh. He arrived in Exeter in 1963 with a passion for paediatric orthopaedics and he hoped to open a centre for scoliosis. However, he found three surgeons already specialising in this area and had to be content with traumatology.[93] The first arthroplasty of the hip in Exeter followed in April 1965.

The Exeter team had several reasons for developing a new THR. The McKee-Farrar prosthesis was used in Exeter from 1965–69 and many problems with loosening were encountered. The surgeons could not get the prostheses for the Charnley low-friction arthroplasty because three out of the four surgeons in Exeter used the posterior surgical approach, and they could not see any benefit in changing to Charnley's more complex method and taking a training course at Wrightington. The problems experienced with the McKee-Farrar prosthesis meant that they did not want to risk using another metal on metal bearing like the Ring prosthesis.[94] So they decided to adapt the Charnley to their own surgical approach, thinking that other surgeons might well share their preferences.

Research for what was originally called the Ling-Lee hip began in 1967 (before the fashion for calling hips after place names rather than creators). Lee and Ling worked as equal partners in a team with both names appearing on the patents that were subsequently applied for. Although Ling took overall responsibility for the hip, it was agreed that Lee would receive payment from the manufacturers for his work. This arrangement continues to date and Howmedica have supported Lee's post at the University of Exeter since the 1970s.[95]

The inspiration was the mature Charnley prosthesis, but the evolution of the prosthesis and surgical instruments was guided by the specific requirements of the posterior approach to the hip. The initial design incorporated many original features. The acetabular cup was offset to allow a thick layer of HDPE. The femoral component was collarless, with a double taper to ensure correct insertion and the extrusion of bone cement. The choice of femoral head size of 29mm was a compromise between the Charnley and McKee-Farrar.[96] The stem of the femoral component was also offset to best mimic the biomechanics of the normal hip. The material of the femoral component was polished stainless steel, as with the original Charnley prosthesis.

The first clinical experience, from 1970 to 1975, was reported in 1978.[97] There were subsidence problems within 18 months; several stems fractured and the entire original batch had a sharp corner on the femoral prosthesis, which increased subsidence as the corner cut into the cement. That was a problem with the original design, but in addition, 95 of the first 443 stems had come back from the manufacturer with reduced neck dimensions, changed for reasons of symmetry and ease of manufacture. That was a reason for some of the fractures, and it emphasises that the separation of the local team and the manufacturer was greater than in the Charnley case.[98]

After these initial troubles, a greater range of stem sizes were introduced and new hips were manufactured in a stronger stainless steel (316L). Because British Standards for surgical implantation did not require this new steel to be polished, Howmedica decided to cut the cost by issuing the stems in matt form. There were hardly any fractures with this stronger material, but there was a marked increase in incidences of aseptic loosening and bone loss (lysis).[99] After extensive research over ten years, it was finally ascertained that the matt finish was largely responsible; an apparently minor difference in the finishing of the metal surfaces had led to major problems. The polished stem was reintroduced in 1986.

A device called a 'centraliser' was placed on the bottom of the femoral stem to ensure that the stem went down the centre of the medullary canal and that equal amounts of cement were between the bone and the stem on all sides of the stem. A high-pressure cement gun was also developed at Exeter to give even distribution of cement around the stem. These two developments helped compensate for the limited visibility using the posterior surgical approach; it is difficult to see directly down the medullary canal from the posterior aspect. Charnley used an elaborate surgical technique to get an open field of view; the Exeter approach limited the visibility, so they compensated by devices which helped ensure correct positioning and cementation – a good example of trade-offs.

In 1988 a modular hip was introduced with interchangeable heads, to get the best combination of head and stem for a particular patient and to simplify the process of revision surgery. The offset cup and head were dropped in favour of a simpler design but the double taper proved very successful and was adapted for many other prostheses (including the triple-tapered C-Stem by DePuy). The advanced cementing techniques developed at Exeter were used extensively at other hip centres. By the end of the century, the Exeter had become the most popular of the British models.[100]

Reflections on surgical inventions

This chapter has traced the development of five different prostheses in their local contexts. Each local context was usually a 'semi-closed system'; the local research was kept secret until the long-term results were known, and there was little contact with other hip centres apart from through publications, which were delayed. The contact that existed was mostly on a personal level, like Charnley's friendships with Müller and Judet and the annual informal meetings of the British Orthopaedic Ski Group (of which Charnley and McKee were members).[101]

In all cases the development was surgeon-led, but engineers of various kinds were variously involved. McKee in Norwich did most of the developmental work himself, with the occasional consultation with Norwich City engineers. At Wrightington, Charnley relied on the ingenuity of a technician-engineer to solve problems, working to very basic sketches and ideas, and making *ad hoc* modifications as they went along. There and at Redhill, the surgeon essentially employed the engineer and they had informal discussions from day-to-day. At Stanmore and Exeter, with more sophisticated engineering input, designs were negotiated, with more complex briefs and a more systematic programme of biomaterials research; that the engineers involved held university posts gave them distance from the hospital context. Indeed, at Stanmore, ten years before Exeter, the surgeon-engineer (Scales) took a lead role, publishing widely and doing much to create the professional role of biomedical engineer.

There were great differences in the designs and tools, but very similar construction materials were used. After initial differences over the question of whether a metal on metal or a metal on plastic articulation should be used, by the mid-1970s they were nearly all using metal on plastic articulations. The exception was Ring, who held out until the late 1970s before switching to plastic cups; and then began to experiment with Titanium in the 1980s.

The surgical approach proved a very important factor in determining the development of the prosthesis. In Norwich McKee used the anterolateral approach and found it initially difficult to position the cup at the correct angle of incidence. The lateral approach (used at Wrightington and Stanmore) gave good visibility but the removal and wiring back of the greater trochanter increased operation times and required considerable skill. The posterior approach (as used at Exeter and Redhill) led to the development of special tools to ensure the correct positioning of the prosthetic components.

As we shall further see in the next chapter, the relationships with manufacturers changed as hip prostheses shifted from a cottage industry to a multi-million pound business. Wrightington was able to impose tight control of the product with monthly inspections by Craven, and Charnley kept control over whom the prostheses could be sold to. Ten years later, Exeter had much less control over the final product; changes of design and materials were sometimes imposed at manufacturing level. Legal contracts and disclaimers were drawn up immediately.

Most British orthopaedic surgeons learned THR from one of these centres. In this respect Stanmore and Wrightington were advantaged – the former as a national special hospital, the latter as a unit devoted to joint replacements which served as a training centre for the Northwest, as well as for foreign surgeons. Charnley had protected the disseminations of his procedure, at least in Britain, and that too had been advantageous, by comparison with McKee's early work which was imitated badly. Exeter learned from the others, both in designs and in setting up courses.

But even for the later devices, one sees the sequential improvisation which characterised all these programmes. Preliminary testing was minimal and animal work was rare; the surgeons moved rapidly to test materials and designs in patients. All this was characteristic of surgical developments from the later 19th century, but contrasts strongly with the systematic trial of drugs that came to be required in the late 1960s, and which were extended to hips prostheses in the 1990s.

In the next chapter we will explore in more detail the surgeon-company relations in Charnley's programme with Thackray – after we have said a little more about the other British companies.

3
The Industrial Development of British Hip Replacements and the Charnley/Thackray Relationship

Introduction

In the last chapter we focussed on surgeons and engineers, but we also mentioned the companies with which they worked. Deloro Stellite, the metal company, sold the McKee-Farrar and initially made some of the Stanmore models; Zimmer UK, a major orthopaedic supplier, was mainly responsible for the Stanmore; and the London Splint Company worked with the Exeter team, and was taken over by their long-time American associate, Howmedica. We introduce the present chapter by briefly exploring the histories of these British companies, before we go on to detail the relationship of Thackray with Charnley. This introduction will also show how the THR business built on the single-component hip prostheses which had been invented in the US but were also manufactured and sold in the UK.

Victorian Britain was famous the world over, not just for large factories but for small companies that prided themselves on high levels of craftsmanship. One such was Down Bros., the largest of the British companies which pioneered the manufacture of artificial hips and produced catalogues in which we can follow the development of the technology. The company began with the work of a craftsman, Millikin, who was appointed as instrument maker to Guy's Hospital in London, and who then established his own small company. In 1879, Millikin was persuaded into a partnership with Arthur W. Down, and the resultant company, Millikin and Down, continued to supply splints and instruments to Guy's and other London hospitals.[1] By 1881 Arthur Down had bought out Millikin and brought his brother into the business, christening it Down Bros. They built their first factory in 1892, moving to larger premises in 1908, and again in 1918.[2] By 1935, the company had offices in central London, where it also had a show room and fitting room on New Cavendish Street, near many of the major hospitals in London (and Harley Street, the centre of expensive private practice). In 1946, Down Bros. bought Mayer and Phelps, who were famous for their ear, nose and throat instrumentation, and the company continued to expand. By 1951, Down Bros. were operating three factories: two in London and one in Brimsdown.[3]

A Down Bros. catalogue from 1951 showed a range of orthopaedic products that had been designed in the US and Europe. It included Stryker beds designed by an American surgeon, and orthopaedic instruments designed and made by Codman and Shurtleff in Boston for such noted surgeons as Otto Aufranc.[4] The catalogue from 1951 carried a Judet hip in 12 standard sizes, as well as nine different Smith-Petersen moulds for cup arthroplasty.[5] In addition, there were two more Perspex endoprostheses that were modifications of the original Judet – one by St John Dudley Buxton with an 'umbrella frame X-ray indicator', and another by K.I. Nissen who added an X-ray marker in the shape of a nail down the centre of the prosthesis. A stainless steel endoprosthesis, which screwed into the femur, designed by Norman S. Martin from Belfast, was also featured.[6] Such catalogues make two important points: that from early in the century there was a considerable American presence, and that besides the well-known, enduring designs of endoprostheses, many early alternatives were devised by individual surgeons and widely advertised.

As the 1960s progressed, more styles of prostheses were added to the range.[7] The 1961 catalogue included the Moore and the Thompson femoral prostheses; as we saw earlier, both were designed by American surgeons. The Moore was manufactured in stainless steel and in titanium, and the Thompson was made from titanium. Among the British pioneers of THR, Down Surgical were the manufacturers of the Redhill hip, but as we have seen, they lost-out to Zimmer UK when more investment was required.

Another firm that that became involved in the prosthesis market in Britain from the early stages was the London Splint Company. One of their catalogues from 1964 carried a total of nine different femoral endoprostheses, including the Matchett Brown, the Wittebol hip from Amsterdam, the Eicher Hip (Eicher worked with Zimmer USA originally on one of their first prostheses), the Jergensen from San Francisco and the McBride from Oklahoma City.[8] The London Splint Company also advertised the Austin Moore and the Thompson, which were the more standard prostheses in the early 1960s. So while the London Splint

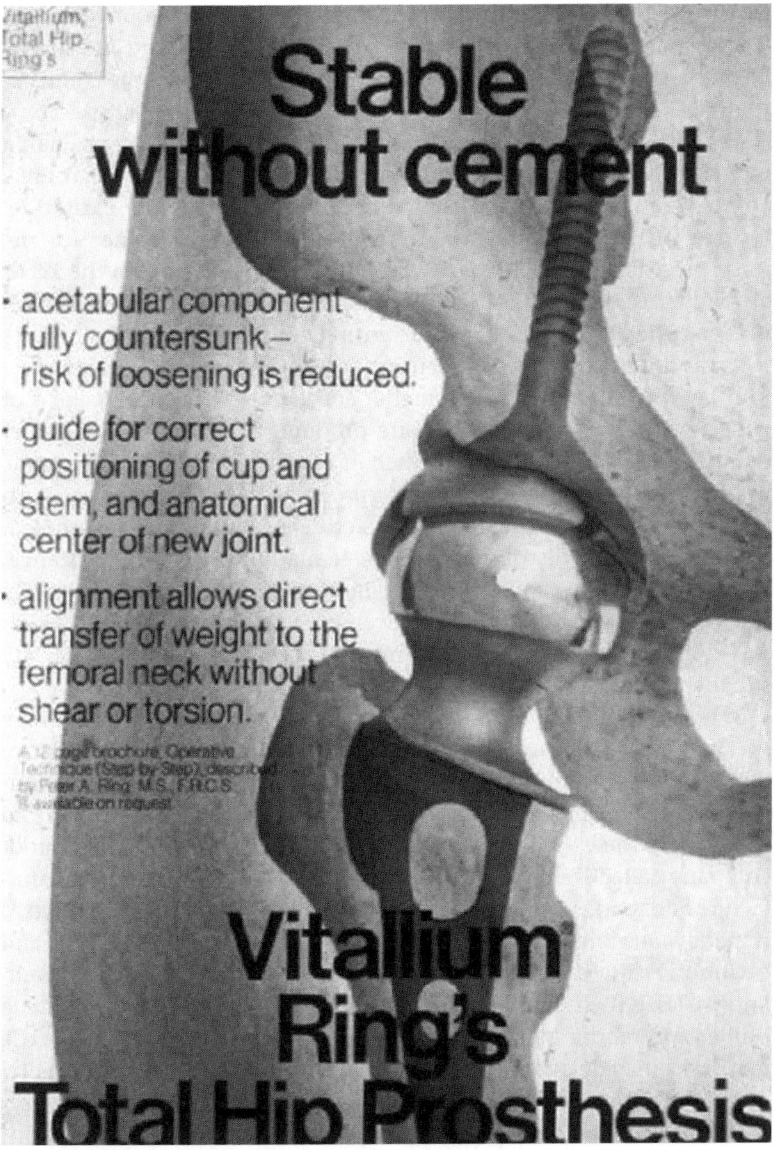

3.1 Down Surgical Publicity Poster for the Ring THR

Company was British, the products it advertised were drawn from an international range of manufacturers and surgeons. Indeed it was closely associated with Howmedica USA, and in 1972 changed its name to Howmedica (UK) Limited, claiming that the old name was misleading when it had become involved in the 'manufacture and supply of specialised surgical implants and appliances'.[9] This was the company that developed the Exeter hip.

As the business of hip prostheses and the demands for associated medical equipment grew, more companies chose to enter that area. Some were specialists in particular materials. For example, Lusterlite, located in Leeds, manufactured acrylic heads under licence for the Judet prosthesis in the 1950s.[10] A pamphlet from Lusterlite in the 1960s also advertised two titanium Thompson stems, one with their trademark acrylic head, the other made entirely of titanium.[11]

It was not uncommon for licensing agreements to be shared among a number of companies, or for smaller companies with specific areas of expertise to combine to make one product. Other small companies manufactured under licence on behalf of a larger entity whose name was used on the product.[12] Specialist products such as cement were originally the province of smaller manufacturers, but many were absorbed, or their products distributed, by larger manufacturers. For instance, North Hill Plastics, which made cement, was bought by Howmedica to gain ownership of the Simplex cement. The CMW cement, which was closely associated with Thackray and the Charnley prosthesis, was supplied to surgeons through Downs and Deloro Surgical.[13]

In both Britain and then the US, some early hip manufacturers were specialists in sophisticated alloys, rather than instrument or appliance makers. Deloro Stellite, based in Swindon, Wiltshire, was a manufacturer of chrome-cobalt products from the late 1950s, similar to Howmet in the US (discussed in Chapter 4); it established an offshoot company called Deloro Surgical (UK) Ltd in 1972 to specifically manufacture and market orthopaedic prostheses and other surgical products.[14] It made Stanmore hips in the early 1970s on behalf of Zimmer UK and became a major distributor of THRs; by 1974, it advertised eight well-known THRs including a Charnley-type, a McKee-Farrar, a Stanmore, a Müller-type and a McKee-Farrar-Arden, all manufactured in its trade mark chrome-cobalt alloy Vinertia.[15] Deloro Surgical was taken over by Zimmer UK in 1980; but again this is a transatlantic tale.

Zimmer UK had been registered in 1964 in Britain after a notorious row between Justin O. Zimmer, who established his company in 1927 in Warsaw, Indiana, and one of his salesmen, who then went to Europe and established the Zimmer name in many countries.[16] With offices in Glamorgan, Zimmer UK was famous for the frames, but it also developed the Stanmore prosthesis, and it gained the Ring project.

One British company was established to import the Müller prosthesis in the late 1960s. Joint Replacement Instrumentation was set up by Ronald Furlong who, at that time, was Director of Orthopaedics at St Thomas' Hospital in London. As none of the other companies was willing to import the Müller prosthesis, Furlong was granted a licence to import them from Switzerland.[17] The prosthesis was cast in the UK at Centaur Cast Alloys in Sheffield by 1975, and by 1977 JRI (Manufacturing) Ltd was established in Sheffield, with one employee with a finishing machine that he had designed and made himself. Other companies that entered into orthopaedic manufacture in the 1950s included Howse, based in London, who made a range of prostheses including the Howse-Arden, McKee-Farrar and the Monk.[18] Throughout the 1970s and early 1980s Howse advertised itself as a 'completely British company', stressing its independence from the global mergers and international collaborations that increased throughout these decades. Howse was taken over by Zimmer UK in the late 1980s.

By the 1970s, Britain had a substantial medical export business. Although the US was the largest supplier of high technology medical devices in the world, British companies also occupied an important part of the market. Britain was the third largest exporter of medical equipment in the 1970s, after the US and Germany, exporting approximately £35 million worth of different products.[19] The British Hospitals Export Council worked with companies so they could sell their products overseas; the business was new and the arrangements often informal. One representative from the industry recalled that sometimes designs came on the back of an envelope or a cigarette packet.[20] The CT scanner was one of these new British products, and we shall return to this export business in our last chapter.

But as John Charnley developed the most successful of the two-component hip replacements, he chose to work, not with one of the established manufacturers of prostheses or a new metal company, but with his favourite maker of surgical instruments. And for this collaboration we have sufficient documentation to conduct a detailed exploration.

John Charnley and Thackray

The company Chas F. Thackray[21] began when Charles Thackray and Scurrah Wainwright bought a pharmacy in Leeds in 1902. The Great War was good for the medical business and they grew substantially, manufacturing items such as instruments, sterilisers, operating tables and the Washington Haigh field dressing. After the war they maintained a reputation as a quality maker of instruments.[22]

Earlier in his career, Charnley had obtained instruments from Down Bros. and Mayer and Phelps of London,[23] but he had worked with Thackray from 1947 and appreciated their expertise and craftsmanship.[24] By the late 1950s, when Down Bros. already carried a number of prostheses, including American-designed products, Charnley may have felt that a company appointed as sole manufacturer would be more likely to attend to his wishes. Moreover, Thackray's base in Leeds was within a reasonable travelling distance from Wrightington. It was a large enough business to invest in a speculative technology, but small enough to be concerned with craftsmanship.

We have seen in the previous chapter that much of the early work was done by Harry Craven, Charnley's technician at Wrightington. Charnley had tried to work with graduate engineers but found it frustrating. His views were made plain early in 1961, when the Wrightington workshop was newly funded and a note appeared in *The Lancet* calling for a closer alliance between surgeons and graduate engineers, and for national funding for innovations and collaboration. It concluded that the initiative 'must come from surgeons who are far-sighted enough to appreciate the full resources of present-day mechanical engineering, metallurgy, and electronics'.[25] Charnley wrote back criticising the article; he was fed up with the cool reception he had got for his projects from academic engineering departments, who were only ever prepared to give spare-time help 'from a junior lecturer whose work is overshadowed by the preparation of his own thesis. The tendency is for the engineer to feel himself to be a "stooge" for the surgeon'.[26] The field did not offer a career for a first-class engineer; there was not enough money for their services in hospitals. Charnley did not add that he had got around this problem by hiring his engineer from industry.

Even when Thackray was making and selling the hips, Craven and Charnley remained important in the process of production. They made a great team, partly because Craven knew how to deal with Charnley's fiery temper. Until 1968, the heads of the femoral components were given a final polish and inspection at Wrightington, and a check was made there of all instruments. Even when production had increased substantially, 5% of instruments and implants were sent to Wrightington for inspection.

The plastic sockets were initially made at Wrightington, and took about 45 minutes to fashion by hand with the model-making equipment at Craven's disposal. As demand increased, and in his characteristic northern style, Charnley asked Craven to design a machine to manufacture the plastic cups: 'You'd better come up with something or you can go up the road.' With the help of his newly-appointed assistant, Geoff Middleton, but little money for raw materials, Craven got most of what he needed from a local scrap yard. The resulting machine featured a special chuck to clamp the polyethylene and various tools that worked the material to form the cup. A motorbike gearbox was added to control the speed! Charnley was very impressed even though Craven and Middleton mischievously told him that it took seven minutes to make a cup when it could actually be done in three.

A discussion followed about what was to be done with the machine and, as Craven had designed and made it, Charnley agreed that he could take it home and make the cups in his spare time. However, the next day Charnley had changed his mind: he didn't want to appear stingy. Since Craven had made it in hospital time, it belonged to the hospital. It would be sold to Thackray, and Craven would reap the profits. Thackray then began manufacturing both the cups and the femoral prosthesis and standardised the procedure. The developmental stage was complete.

Charnley's growing reputation remained 'an important selling point',[27] but he waived the rights to royalties; instead, £1 would be donated to his research fund for every prosthesis sold. Craven moved on to help Charnley in developing the innovations of the clean-air operating

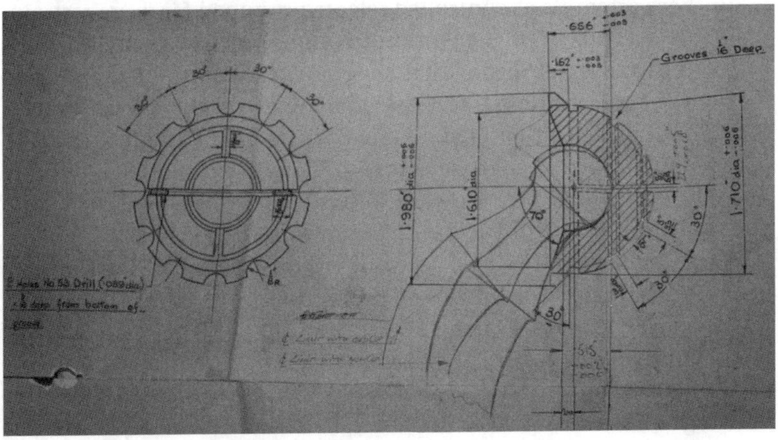

3.2 Thackray Drawings of the Charnley Femoral Prosthesis and Acetabular Cup in the mid-1960s

system and the design of surgical instruments and the tray system, specifically tailored to the operation. The clean-air equipment, which we discuss in Chapter 5, was to further strengthen Charnley's reputation as the principal developer of THR; but it was much less copied than the THR. Charnley's tray system for surgical instruments had to compete with a system simultaneously developed in Edinburgh, and the Scottish version, with superior sterilising techniques, was used much more widely. After Craven left in 1966, Thackray also agreed to pay the salaries of two skilled technicians who deserved a higher rate than their NHS classification as 'orthopaedic appliance makers' could command.[28] And thus in a roundabout way Craven's input was recognised.

The requests from surgeons to both Charnley and Thackray had increased rapidly from 1960, when Charnley published his article in the BMJ on the surgery of the hip joint,[29] followed by another in *The Lancet* in 1961. But mindful of the disaster with the Teflon sockets, Charnley, as we noted, was wary about releasing his new prosthesis on the open market. Thackray's advertisements stated that while six-year results were available, a ten-year study would be preferable, so 'in compliance with Mr Charnley's wishes, the implants are available only to surgeons who have visited Wrightington'.[30] By 1968, 6,000 implants were being manufactured per year but quantities were still insufficient to match demand.[31] Implants and instruments had to be locked up at the factory, and access limited, to ensure that requests were filled in the correct order.[32]

But these attempts at controlled release presumed far more restraint than was ever likely in the world of competitive surgeons, let alone that of businesses. It meant that surgeons were more likely to go to their own manufacturers and ask for copies of the Charnley hip. Though Charnley attempted to maintain control of the diffusion process, the implant purchase course at Wrightington was reduced to two days as the 1960s progressed, and by 1968 was abandoned completely. Yet the foreword of the 1975 Thackray catalogue still claimed 'The implants should be available only to men who had worked with me at Wrightington for a period of six months or more.'[33]

Charnley also tried to control diffusion by limiting the release of images of his prosthesis. The first catalogue in 1968 showed only the instruments and the tray system. By 1970, simple single-aspect line drawings illustrated two prostheses, two acetabular cups and the cement restrictor.[34] There was a basic description but no information about what it was fashioned from except 'stainless steel'.[35] These simple

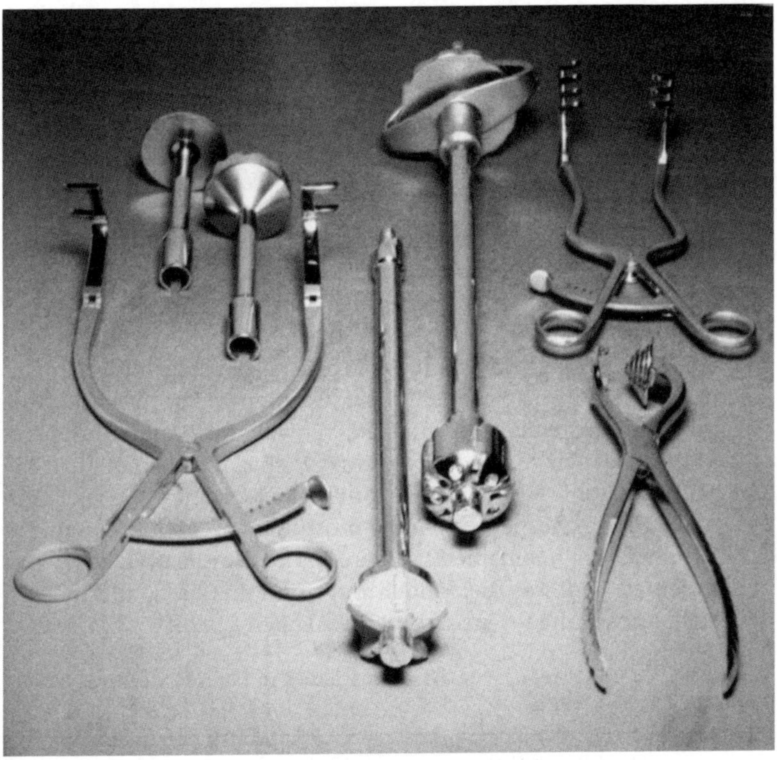

3.3 Surgical Instruments designed by John Charnley and made by Chas F. Thackray

line drawings also appeared in advertisements in the *Journal of Bone and Joint Surgery*, although these ads also gave information about testing standards and the stainless steel's compliance with British Standard Specifications.[36] Charnley feared 'irresponsible modifications' by other manufacturers attempting to copy his design from photographs.[37] Later catalogues from the late 1970s and early 1980s contained photographs from several angles and a clear description of the prosthesis. For example, the 1981 Thackray catalogue contained 21 femoral stems and 11 acetabular cups with photographs and descriptions.[38] But by then the prosthesis had long been copied and altered.

Thackray acquiesced in Charnley's decision regarding both the restricted release and lack of detail in the catalogue. As Charnley reported in the 1975 catalogue, Thackray had 'begged to know when I would lift my restrictions'.[39] He continued, 'one can imagine the pressures and temptation I should have been subjected to had Thackrays' orientation been purely economical'.[40] And yet, because production was on a small scale, Charnley often wrote to them complaining that orders for instruments and implants to approved surgeons were not being filled.[41]

He insisted that Thackray keep the price of the hip-replacement components as low as possible – in 1968 the Charnley hip cost £17 10s 0d.[42] Part of his reason was to make the procedure more available on the NHS. Another was to minimise competition, as any rival manufacturer would not find it cost effective to break into the market.[43] But this assumed that the Charnley hips were widely available, while he had restricted supply for good reason. And to American surgeons and manufacturers, the low price seemed ridiculous compared to the cost of the operation and hospitalisation. Indeed, allowing for product improvements, the Charnley hip did stay relatively cheap in the UK until the late 1990s (see Table 3.1). But if we can briefly look ahead: in the US in 1996, though the cheapest cemented prosthesis cost about the same as the Charnley, which remained standard in the UK, the average US cost of a cemented stem was then about four times that level, and the cost of coated, uncemented stems was twice as much again.[44]

Table 3.1 Unit Cost of Charley Hip Prostheses, in Pounds Sterling, 1968–1974[45]

	Actual Prices (£s)	2005 Prices (£s)	Index
1968	17.5	224	100
1995	285	362	162
1996	306	380	170
2004	780	795	355

Once the prosthesis was released to a few surgeons, any effort to stem the flood of copies was near useless. Charnley wrote letters to Thackray detailing his worries about the expansion of other medical device manufacturers' practices such as the wax casting of his instruments and prostheses. Thackray advertised in the *Journal of Bone and Joint Surgery* that there was 'no need to buy copies...the authentic Charnley is available to all!'[46] But by 1975, at least six other manufacturers were advertising 'Charnley-type' prostheses in that journal.[47]

Unauthorized copies affected Thackray's market share. In order to sell more prostheses, Thackray would take samples and testing-rigs to industry shows in order to demonstrate the low frictional properties of the components. Although these demonstrations were a success, many manufacturers were still more likely to make copies for surgeons as opposed to ordering them from Thackray. One manufacturer did ask Thackray if they could expand the range of the prosthesis by manufacturing it in chrome-cobalt, but Charnley declined to give his permission.[48] Thackray tried to persuade him that since they did not hold a patent in the US, the company would lose revenue on sales if another firm manufactured copy prostheses in Vitallium.[49]

Charnley suffered the indignity of the *Journal of Bone and Joint Surgery* displaying advertisements of his tools and implants under other manufacturers and surgeons' names.[50] One company copied an acetabular cup holder that was marked differently from the one made by Thackray, and when this cup holder was used the component was inserted at the wrong angle. Charnley then noted to Thackray, 'I think some important use can be made for this in ridiculing the Americans and in putting the point about the importance of patents not being cribbed.'[51]

Charnley wrote revealingly of American industry that it 'is always excited at a new development in a foreign country but very soon forgets its source of origin and takes it over completely'.[52] In 1971, Thackray and Charnley had opted for a sole distributor, Codman and Shurtleff Inc. of Randolph, Massachusetts to represent them in the US,[53] but the damage had been done and the prosthesis did not increase its market share. Whilst Thackray continued to advertise in the British version of the *Journal of Bone and Joint Surgery*, Codman, or Cintor Orthopaedic Division of Codman as it became known by 1975, advertised Charnley products in the American version.

In the 1960s, in the initial excitement of the low friction plastic and stainless steel combination, the Charnley name was central. He received numerous requests from companies wanting to advertise their designs as authorised by him.[54] Generally he refused, because he did not have a sufficient time to test each prosthesis. The one exception was for his Swiss friend Maurice Müller – the surgeon behind the AO company, though Müller's hip designs were marketed separately. In fact, this Charnley-Müller design reverted to the larger femoral head that Charnley had discarded when he became concerned with the coefficient of friction, even though smaller heads risked more dislocations.[55]

Charnley appears to have been uninterested in making money from his prostheses. He worried about the success of the operation and how it could be implemented in the NHS, rather than protecting his innovations by patents. He came to think this an error. After 1969, with many copies on the market, both Charnley and Thackray were more careful about the necessity for patenting and ensured that licences were taken out, sometimes long after the design was in general use.[56] Thackray itself held a range of patents, from the process for finishing the heads of the femoral components to the tiny cement restrictor which prevents cement from entering into the medullary canal.[57] Codman listed the US patent numbers in all their journal advertising from 1972, though Charnley believed that his American agents were 'basically against helping with patents from foreign countries'. The controversy over international patents continued throughout Charnley's life.[58]

By the mid-1970s, part of the reason for Charnley's new attentiveness was personal. At the urging of Thackray, Charnley formed a company to collect royalties at a lower rate of tax.[59] Charnley's protective attitude towards his prosthesis continued until his death in 1982, and questions of property were also important for his colleagues and successors at Wrightington Hospital.

Though Charnley had resented the autocratic style of his Manchester mentor, Harry Platt, and had found himself in rivalry with Platt's other protégé, he got himself into a similar situation at Wrightington, where Kevin Hardinge concentrated on the clinical work and Michael Wroblewski on research. Hardinge felt Charnley's style shortly after his appointment, when he developed a cement plug to stop cement migrating down the medullary canal away from the stem of the prosthesis. Thackray refused to take it lest they offend Charnley, whose rival design had never worked. So Hardinge took his 'shuttle cock' design to Howmedica, and it worked and sold very well.

After this experience and again with Howmedica, Hardinge developed the Wrightington Total Hip Replacement in the early 1980s. This had a specially-shaped stem and was also designed to be used with a novel surgical approach – the direct lateral, which intersected the muscles above the great trochanter rather than removing this ridge of bone. The approach retained most of the visibility of lateral approach but shortened the length of the operation and patient recovery time.[60] Although it was a minor modification to the Charnley concept, he would have never sanctioned it. After Charnley's death in 1982, Wroblewski developed a new cup design and a 'C' stem.[61]

As the Hardinge case shows, and as we discussed in the last chapter for the Exeter hip, issues of surgical approach and techniques were often crucial. In 1972 Thackray itself designed a prosthesis with a longer neck for surgeons who did not want to remove the greater trochanter.[62] Some American surgeons were concerned about breakage of the wires that lashed the greater trochanter back on to the femur; most sought alternative methods, but William Harris, in Boston, introduced a device to tighten the wires.[63] Charnley himself continued to develop his original operation, spending huge amounts of time on refining the trochanter removal and reattachment that many surgeons thought unnecessary.[64]

Materials, too, came under scrutiny and new manufacturers adopted new alternatives, particularly in metals. Other surgeons, particularly Americans, suggested that Charnley's stainless steel could not compare with the chrome-cobalt alloy which went by the commercial name of Vitallium and was already used for femoral prostheses. As we shall discuss in Chapter 5, the controversy raged on both sides of the Atlantic.[65] Charnley and Thackray had an uphill battle to persuade surgeons of the benefits of stainless steel, though it was significantly cheaper than Vitallium and titanium both in its raw and production costs. As Charnley said in a letter,

> The NHS in Britain, for instance, will take stainless steel prostheses if somebody such as myself will back them as being completely reliable...The situation where chrome-cobalt is demanded relates to standards of living which are quite out of comparison with this country.[66]

And as Charnley later wrote to Thackray, the company might as well benefit from the lack of price constraints in the US:

> There is absolutely no doubt that surgeons in the USA have no idea of how much implants cost...This is because these are paid for by insurance companies. It seems, therefore, to me that the price to which your agents knock them up is not particularly important provided that patent cover can be carried to guarantee that they are made by you or made by firms in America licensed to make them for you.[67]

In 1971 Charnley finally allowed Thackray to make his prosthesis in Vitallium – though he never abandoned stainless steel. In a letter to Thackray he spoke of 'perfectionist engineers who are interested in metals, but have no experience of surgical appliances'.[68] To cast their products in chrome-cobalt, Thackray used Aeromed, a Sheffield company they half owned.[69] In 1975, in a complete turnaround, Charnley wrote to Thackray suggesting that an American company should take over the manufacture of the chrome-cobalt prostheses.[70]

In the use of stainless steel, Charnley had government backing: the Department of Health and Social Security (DHSS) had officially agreed that stainless steel was their preferred implant material as opposed to Vitallium, probably because of its lower cost and the long-term success of the Charnley steel implants. In 1971 Charnley thought that the DHSS should publish this evidence 'in the interests of the export trade and British trade in general'.[71] But criticism and problems resulting in potential litigation, mainly from the US, had already caused Thackray to try to enhance the quality of their stainless steel. By 1971, an improved type of stainless steel called 316LVM was being used and quality control methods had been put in place by a consultant metallurgist appointed by Thackray.

In 1974, as a result of a fracture of a Charnley prosthesis, Thackray was threatened with litigation in the US by two patients.[72] The report that was returned from the Thackray metallurgist was that the breakage was due to corrosion fatigue, which confirmed to the Americans that stainless steel was inferior to chrome-cobalt.[73] Charnley stated in a letter to Thackray that it had taken 6,000 operations before this problem had become obvious, blamed the surgical technique, and also pointed out that other companies had to change their designs when problems became apparent.[74] He also extended the blame to the DHSS.

> If the generation of failed prostheses which are under criticism was from this era then I think the DHSS has some responsibility here in that it very definitely influenced my opinion and did nothing to make me look elsewhere for better materials.[75]

But in the same year, J.T. Scales, who had designed the Stanmore hip, compared prostheses manufactured from chrome-cobalt, stainless steel and titanium and found that his stainless steel cold-formed prosthesis showed a higher tensile and proof stress.[76] Thackray then began investigating this cold-forming process which also allowed flexibility in design so that alterations, which Charnley still made regularly, could be accommodated without increasing costs.[77] Improvements continued to be made as Thackray experimented with other compounds of stainless steel and at the end of the decade a much superior type, called Ortron, was released on the market.

Through the 1970s, Thackray had taken more responsibility for design and research, hiring new specialised personnel and using other research facilities such as universities to test and perfect new materials. They expanded their premises, adding a whole new wing in Beeston to accommodate production of the hip prosthesis. Different prostheses introduced during the 1970s included the round-back in 1974 and in 1975 the flanged-back which reduced subsidence. This new prosthesis was designed in two weights, thus meeting the need for a device for the heavier male patient.

Other considerations became obvious as time went on. As younger patients underwent the procedure and the prosthesis could no longer be guaranteed to last the lifetime of the patient, the need for a revision prosthesis became evident. Charnley designed one and Thackray were manufacturing it by 1975. This brought up the question of prosthesis removal, so models were designed which could be removed more easily. When certain deficiencies were exposed, for example that the stainless steel prosthesis was subject to pitting, a coating called vaquasheen was applied to the prostheses that limited the effect. While the original theory behind the Charnley design remained successful, new innovations were constantly on the drawing board.

Charnley was a perfectionist when it came to design and materials, and constantly made alterations to surgical instruments and prostheses. The engineers at Wrightington did most of the research and development; Thackray was given alterations to designs and would make the necessary changes. And Charnley would get impatient if they waited to sell their current stocks before implementing new designs. In 1971 he wrote,

> Where new patterns have been designed for better surgical efficiency, and to avoid errors inherent in old patterns, then obviously a radical scrapping should be adopted.[78]

Letters detailing new instrumentation and alterations to designs streamed steadily between Charnley's office at Wrightington and Thackray head office, and the relationship was not always congenial. More than once Charnley threatened to take his new flanged design to another manufacturer. In 1974, Charnley had to be reminded that Thackray was the sole authorised manufacturer, and their American distributor could sue Charnley if a patented device were offered to another company.[79]

One part of Charnley appreciated the readiness of patients and companies to follow doctor's orders, without complaining, pestering or worrying unduly about money. But another part of him was irritated by complacency and saw the advantages of demanding patients and keen market competition. Sometimes he saw Thackray as Victorian and complacent; it would be left behind in the rapidly expanding market for medical implants. The 'British have no concept of the standards of quality required on the Continent or in North America'.[80] He had liked the freedom which the NHS had given him, to focus on a technical and professional project, but as the 1970s wore on he grew impatient with Britain and its governments. In this, he followed his political heroine – Margaret Thatcher.[81]

After Charnley: Manufacturers in Britain after 1982

Sir John Charnley died in 1982, and one of the closest relationships between surgeon and manufacturer ended. Thackray continued to make Charnley hip prostheses and branched out into new materials such as ceramics. Thackray finally succumbed to the global domination of the orthopaedics industry in 1990 when it was purchased by DePuy's parent company at the time, Boehringer Mannheim. Thus DePuy added cement fixation to their previous expertise in biological fixation.

Zimmer UK took over Deloro Surgical in 1980, calling the new company Zimmer Deloro Surgical, but in the mid-1980s the name reverted to Zimmer Orthopaedic when the company was bought by the Orthopaedic Equipment Company (OEC), from the same town as Zimmer US; OEC was bought by Biomet in 1985.[82] We see here the ways in which a British orthopaedic company (albeit with American roots) first used and then took over a British specialist metal company, and then was bought up by a major medical company from Warsaw, Indiana, which in turn was bought by one of its neighbours in this world-centre of orthopaedic technology.

Down Bros. which had manufactured the Redhill hip, and supplied it under licence to Howmedica of New Jersey in the 1970s, was taken over by the German company Aesculap in 1988 after a complex series of takeover manoeuvres through other British companies from 1984–1988.[83] Aesculap, which designed both cemented and cementless hips in the 1990s, became a division of B. Braun Medical Ltd, an international multi-arm medical company, based in Germany, with a major factory in Sheffield. This old steel town had a number of specialist metal companies interested in medical products, and new ventures such as JRI were established there in the 1980s and 1990s.

But despite the dominance of the multi-nationals through the late 1980s and 1990s, some new firms were established in Britain. The growth in new manufacturers came mainly from developments in niche markets, or smaller companies that had designed a particular product or procedure. Corin was formed in 1985 with headquarters in Cirencester;[84] it manufactured and distributed the Freeman Modular Total Hip System, which was a metal polyethylene combination, but it gained its reputation from its hip-resurfacing product, Cormet. It bought Alphanorm, a German orthopaedic business in 2002, and became large enough to divest itself of OSI which distributed its products in the US and established its own presence there. The late Ronald Furlong designed a prosthesis with the osteoconductive coating hydroxyapatite and it was first used in 1985.

The Joint Replacement Instrumentation Company (JRI) grew strongly in the late 1990s as the demand continued for prostheses that would stimulate bony ingrowth for younger patients. This all-British company manufactured hip implants for the Queen Mother, although as an elderly patient she was given a cemented implant.[85] Other companies included Midland Medical Technologies (MMT) whose principal product was the Birmingham hip-resurfacing procedure, pioneered by two surgeons in the UK. As we have seen, many of these smaller companies were established in order to focus on a particular need, either for a design or a procedure; and in Britain many of them have been surgeon-driven, even in the 1990s.

On a rather different scale and by very different means, another British company entered the hip market. Smith and Nephew did not start out as a manufacturer like Down Bros.: it began in 1856 as a chemist shop in Hull. When the founder's nephew entered into partnership with his uncle in 1896, the shop was called T.J. Smith and Nephew, and they began making medical dressings (this a year after DePuy established itself as a company in the US). WWI brought a massive expansion – from 50 to over 1,200 employees;[86] by 1928 Smith and Nephew had developed Elastoplast band aids, followed by plaster of Paris bandages in 1930.[87] After WWII, Smith and Nephew became a major international company, and in the 1980s chose to focus on high value goods. As we shall see in the next chapter, they soon purchased enough American expertise and capacity to become a world player in THRs.

Reflections on collaborations and companies

This chapter has encompassed two stories; one about Charnley at Wrightington, the other about interactions with the rest of the world and especially the US. The first shows an innovative surgeon developing his programme in a low-status hospital, aided chiefly by a resourceful technician. Charnley seems to have drawn less from other orthopaedic surgeons than from a range of industrial contacts and from colleagues in the dental school at Manchester and the engineering department at UMIST. His work was that of an inventor; it was bricolage. Harry Craven scavenged scrap to build a test rig, or he improvised a 'greenhouse' around the patient's body in the operating theatre to reduce the risk of airborne infections. Charnley put together a stem, a cup, a method of fixation, an operation, a follow-up system, and a germ-free environment. There was very little laboratory work apart from basic histology and pathology, or elaborate calculation apart from simple testing of materials. He decided on a small, metal femoral head, he tried plastics; he developed his technique and systematically varied components to get better results. He began with a dramatic failure, and he recovered to set the standards for the field. He knew how he wanted the field to develop: British surgeons would come to him to learn the methods, which they would then practice safely using the tools and implants from Thackray.

And if Britain had been a surgical island, it might have worked out like that. The other pioneers switched to include variants of his method (as in Norwich and Stanmore), or their programmes were discontinued (as at Redhill), or they began (as in Exeter) with a Charnley variant for surgeons who preferred a different surgical approach and a less fussy operation. In a different world this could have been a tale of Charnley variants, by Thackray and one or two other firms, developing slowly, widening the population base, and refining the materials, the finishing and the designs. Thackray might have developed like AO, as the dominant maker in the field; or if he had used Howmedica, he might have controlled less, but for a wider market.

But the real course was very different. When Charnley's success was clear, imitations followed in Europe and America. Charnley's metal was too cheap and lacked metallurgical sophistication; the cement was a problem because of American drug regulations. Especially in American private markets and cost-unlimited Medicare, there was lots of room for more expensive versions, and for surgeons and companies to try their own designs. None of these was so successful as to displace the Charnley, Stanmore and Exeter from their British home market; in that sense the American varieties were not, in general, better adapted to the bodies of the main groups of patients, when measured by careful follow-up. But they were better adapted to rich markets, competitive surgeons and hospitals, a culture of litigation, and to medical device companies which by the 1980s incorporated at a high level the skills in materials, designs and testing that Charnley had put together *ad hoc* and *pro tem*. Above all, the companies had selling power.

We began this chapter with a range of British companies, some of which already had US linkages. But though the key technology moved westward from Britain, industrial control extended eastward from America. It was the American links which expanded in the UK, and the American companies that eventually took over most of the British market. How the American industry acquired those strengths we shall examine in the next chapter.

4

Medical Technology and Industrial Dynamics: The United States and the Global Market

Introduction

In this chapter we explore the histories of some of the American manufacturers who now dominate the global hip-replacement market. We go on to examine how they have worked with surgeons, not least through technical representatives who visit hospitals and may take a part in the surgical operations. Worldwide, there are 18,000 manufacturers of medical devices, and over 70% of them are small companies with less than 50 employees.[1] Some companies only have one or two specialist product lines, or conduct research for the large manufacturers. However, as in many industries, the large multi-nationals have most of the market share.

From bandages to implants

Orthopaedic firms have been a part of the manufacturing sector in the US since the late 19th century. By the 1940s, many of these companies made splints, nails, bandages and instruments; some of the companies relied on a particular product, as with Stryker's frame.[2] As this chapter will demonstrate, many of the newer manufacturers were offshoots of older companies. The two oldest surgical-device companies in the US, DePuy and Zimmer, were both established by salesmen, in Warsaw, Indiana – which became and remains the home town of the American orthopaedic industry.

Revra DePuy founded DePuy Manufacturing in 1895, the first commercial orthopaedic manufacturer in the US. The first product was a fibre splint to replace the customary wooden ones; the main customer was the US Army.[3] Revra DePuy died in 1921 and his wife continued to manage the business. Justin O. Zimmer, DePuy's first sales representative, tried to buy the company from Depuy's widow, but without success, and in 1927 he formed his own manufacturing premises across the other side of Warsaw. Zimmer's first products were aluminium splints, braces and wound drainage systems. Zimmer was one of the first companies to have a sales booth at the American Academy of Orthopaedic Surgeons (AAOS) meeting – in 1933.[4] After the US entered WWII in 1941, orthopaedic manufacturing grew rapidly. By 1942, Zimmer had 100 employees with sales of US$1 million.

Justin Zimmer started a trend. We have seen that one of Zimmer's salesmen set up an independent company in Europe. Another, Don Richards, left Warsaw in 1934 to start his own manufacturing company, Richards, in Tennessee. In 1950, an employee of Richards established the Wright Manufacturing Company, also in Memphis. And Warsaw also saw new companies, including the OEC (Orthopedic Equipment Company) founded in 1943, and Biomet in 1978. All drew on this small town's large pool of specialist expertise.

In a few cases, surgeons entered the industry. The first such company was started in 1941 by Homer Stryker, an orthopaedic surgeon who had designed the first Stryker bed in the 1930s to help relieve the problem of pressure sores; the beds were widely used in the WWII for servicemen paralysed from spinal injury.[5] Stryker did not manufacture hip-replacement implants until 1979, but they were already making powered instruments, such as the Stryker saw.[6]

Howmedica is one of the older hip manufacturers in the US, but its origins were not those of the traditional orthopaedic manufacturer. Like Deloro Stellite in UK, it derived from a specialist metal company. Originally called Howe Sound, changing its name to Howmet in 1964, it was a metal and casting manufacturer located in New York State and specialising in chrome-cobalt products. The company expanded in the 1960s to take advantage of the growth in chrome-cobalt products, mainly in dentistry. Howmedica was established as a separate manufacturing concern in 1969 specifically to supply the dental and orthopaedic side of the business.

The hip implant market

In the 1950s and early 1960s, single-component hip prostheses became common in the US. Well into the 1960s, when British surgeons were turning to THRs, American surgeons generally preferred *either* a cup or a femoral stem.[7] In a rapidly expanding market, manufacturers used their catalogues to showcase a myriad of endoprostheses, all designed by surgeons from across the US; and as we have seen, many of these models also appeared in British catalogues.[8] Zimmer worked closely with Dr Palmer Eicher who produced their first endoprosthesis in 1950,[9] but he also advertised his prosthesis with the British supplier, Down Bros., which had offices in North America.[10]

The most popular of these hip prostheses were made by several manufacturers and there appeared to be little agreement as to whether any company owned specific patents on particular prostheses. For example, one of the oldest femoral prostheses, the Austin Moore was advertised by Howmedica (at that time named Howmet), Wright Orthopaedic Company, and OEC. While the materials differed, (Howmet and OEC produced theirs in chrome-cobalt and Wright in stainless steel), the design was the same.[11] Richards advertised a different endoprosthesis, the SMO, which was designed by a group of three surgeons, Drs Morrissey, Snyder and Yund, and was released in 1964.[12]

Much of the advertising in the early 1960s in the American editions of the *Journal of Bone and Joint Surgery* was as much about the companies as the products. One advertisement by Zimmer stressed the role of the laboratory in quality control,[13] perhaps in response to claims that their stainless steel was not as good quality as Vitallium. Until the wider development of a greater range of prostheses in the early 1970s, it was the quality of raw materials and manufacture, rather than design innovations, that were key in advertising hip implants.[14]

When they learned of the new two-component hip implants in Britain and in Europe in the 1960s, American companies began to market them. The uptake was not rapid but it was established by the later 1960s, along with various forms of importation. Many companies managed to get around the question of patents by merely copying a design and naming it a 'type'. While by 1972, Wright, Zimmer USA and Howmedica were producing the Müller Type or the Charnley Type, Zimmer produced the Zimmer McKee-Farrar Total Hip, and suggested they had redesigned it.[15] Zimmer also produced its own prostheses, advertised as designed in collaboration with an American surgeon, including their Sbarbaro, a metal on metal articulation which was 'crafted by Zimmer'.[16]

Before 1972, several US companies also entered into agreements with European and British manufacturers to distribute their prostheses in the US. For example, Richards had distribution agreements in the 1970s with Waldemar Link in Hamburg for the St George Bucholz hip.[17] That other European firms entered the American market without distribution agreements with US partners, demonstrated the depth of the market for implants: in the late 1970s and early 1980s, Link prostheses were manufactured in Germany and imported directly to the US.

Already by 1968, DePuy had secured exclusive rights to supply the Müller prosthesis in the US; it was manufactured in Switzerland throughout the 1970s, and continued to be the preferred European-designed prosthesis. Müller's hip was distributed by different supply companies around the world, for example by Hip Replacement Instrumentation in Harley Street, London.[18] Maurice Müller had established his own company, Protek, to manufacture and market hip replacements, but his design, though patented, was much copied and many companies carried this design 'type' through the 1970s. The business of supplying prostheses had become international, and transatlantic deals the norm.

In 1972 the two-component Harris hip became available from Howmedica; made in Vitallium, it was one of the first total hips designed in the US – from information Harris had collected on trips overseas to Britain and from his experience of using two-component prostheses such as the McKee and the Müller.[19] By 1973, American designs included the Bechtol Total Hip System from Richards, the Frost Total Hip from OEC, and Zimmer's Trapezoidal 28. Bechtol of Los Angeles, like Harris of Boston, was a national leader in orthopaedics.

New companies continued to spawn from the more established firms in the 1970s, although with less regularity than in earlier years. In 1978, Biomet was set up by a group of four employees in Warsaw. Like DePuy and Zimmer much earlier, and despite the expertise of one of its founders in engineering design, they established themselves through simple products, including soft-good products such as cervical collars. Then in the early 1980s, Biomet worked closely with two surgeons, Thomas Mallory and William Head to design a new type of acetabular cup (Head) and femoral stem (Mallory). The two surgeons had worked separately for Biomet, but then came together to jointly design a two-component implant. With the release of the Mallory-Head system in 1986, as well as other products including artificial knees, Biomet went from being a very small operation, mainly run by the founders and their wives, to one of the largest manufacturers in the US. When a Dr Tronzo of Florida claimed that the Mallory-Head design appropriated ideas he had previously shared with Biomet, he gained US$20 million in damages in 2001.[20]

Howmedica engineers who had worked with implants, and wished to market their ideas themselves, set up their own company called Osteonics in 1978. In a demonstration of things to come, Osteonics was taken over in 1979 by Stryker. Though Osteonics was then US$240,000 in deficit, it brought Stryker instant expertise in the expanding market for prostheses.[21] In 1987 Zimmer employed 3,500 people with sales of more than US$500 million, but despite the increasing strength of the large manufacturers, small companies continued to come into existence. For example, BioPro was established in 1987 by Charles O. Townley, a surgeon from Michigan who had opened his first orthopaedic clinic in 1950, and who designed his own hip and knee implants.[22] Originally Townley designed the Total Articular Replacement Arthroplasty (TARA) for DePuy,[23] but with BioPro he proved that a surgeon could still market his own implant, albeit with industrial experience.

Relationships with pharmaceuticals

By the early 1970s, American firms were selling large numbers of imported designs and beginning to establish their own models. Over the next decade they bought several of the key European companies and came to dominate the world market. They manufactured European models for their local markets, but they also made them in the US, along with the preferred American variants. They could do so more easily because they had themselves been bought up – by the large pharmaceutical companies which had benefited from the rapid expansion of pharmaceutical sales since the 1940s. The parent companies had excellent connections with hospitals, huge numbers of salespeople, and saw themselves as supplying an increasing range of medical technologies. In 1972, Bristol Myers Squibb bought Zimmer, and Pfizer purchased Howmedica. Boehringer Manneheim followed in 1974 when they bought DePuy through their acquisition of Bio-Dynamics. The process of consolidation with pharmaceuticals continued later when Roche became the parent company of DePuy and Boehringer Manneheim in 1990, and Johnson & Johnson bought DePuy in 1998.

As we noted in the previous chapter, Howmedica bought the London Splint Company in 1972, the year they were themselves taken over by Pfizer. The London Splint Company was the first British medical device company to be taken over by an American company;[24] the two companies had operated very closely since the 1960s, with products made under licence by the London Splint Company for Howmedica.

Howmedica had also established its own companies in continental Europe including Benoist Girard and Co in France.

In 1980, Zimmer purchased Deloro Surgical who manufactured the McKee prosthesis in chrome-cobalt in the UK, which gave them a stronger foothold in the European market.[25] Biomet took over other companies to expand its manufacturing base, and in 1984 it purchased the OEC which had bought Zimmer UK. Biomet acquired Kirschner in 1994 further expanding its market in Europe,[26] and in 1997 they entered into a manufacturing agreement with the pharmaceutical giant Merck.

But this transatlantic process could work in reverse. The British medical manufacturers Smith and Nephew took over Richards in 1986 paying £210 million in order to enter the lucrative medical device market in the US. In 1999 Smith and Nephew increased its number of hip implant designs when it bought some of the lines developed by 3M, which as we shall see later, had experienced difficulties entering a new industry. Smith and Nephew turned its back on plasters, facial cream and feminine hygiene products to concentrate on the more lucrative area of orthopaedic implants and keyhole surgery.[27] It took

4.1 The Birmingham Hip Resurfacing (Femoral & Socket Surfaces)

over various companies, including DePuy's casting and splinting group (2000); and acquired Midland Medical Technologies in the UK, whose principal product was Birmingham Hip Resurfacing, a method of preserving the joints in younger patients who might later need hip replacement surgery.[28] It also explored joint ventures with companies such as Beiersdorf and B. Braun. In 2003 Smith and Nephew had a 7% share of the American prosthesis market and about 25% of the European market, which was higher than any of the big four companies taken individually.[29] Though the parent company was British, the orthopaedics business was run from the US.

As the device companies became larger and richer, their relationships with the pharmaceutical parent-companies were in some cases renegotiated. In 1992 Pfizer made Howmedica independent of its hospital-products group. Perhaps this was to distance its lucrative products such as hip and knee replacements from the less successful part of its hospital group, as in 1991 hundreds of lawsuits were filed against it for defective Bjork-Shiley heart valves which were considered responsible for a number of deaths. By 1998 Pfizer divested itself of Howmedica to Stryker Osteonics. Although hips and knees were considered a profitable part of the business, Pfizer concentrated instead on new drugs. This new orthopaedics company, Stryker-Howmedica, had the largest market share in the late 1990s.

Bristol Myers Squibb listed Zimmer as a separate company on the New York Stock Exchange in 2000, and DePuy is also listed separately from Johnson & Johnson. Biomet, as noted, has an agreement with Merck. So of the American orthopaedic giants, only Stryker-Howmedica is independent of the pharmaceutical business. By 2003, four multinational companies, Biomet, DePuy, Stryker-Howmedica and Zimmer controlled about 85% of the American hip-replacement market.[30]

These American-based companies also dominate the international market. For example, foreign manufacturers control 80% of the US$341 million hip-joint market in Japan, though local manufacturers have attempted to regain control of lost markets.[31] Japanese companies such as Kobe Steel have tried to break into the hip market with prostheses 'specifically designed for the Japanese physique'.[32] Other countries such as Colombia, and countries in the Middle East, do not have the manufacturing infrastructure to support medical devices factories, so they have had to rely on the multi-national companies to supply their hip replacements.[33]

The large manufacturers competed with each other to dominate the market. In 2003, Smith and Nephew engaged in a bidding war with

Zimmer for Sulzer, a European company which had changed its name to Centerpulse when it came to light that one of their manufacturing processes had malfunctioned, leaving many patients with hip implants that were contaminated with machine oil. Eventually the larger company, Zimmer, was the successful bidder and acquired Centerpulse and a larger market share in Europe.[34]

Though the American firms are huge and global, their base remains curiously local. DePuy, and its offspring Zimmer, and its offspring Biomet, are still headquartered in the small mid-western town of Warsaw, Indiana. The implant business has brought higher-bracket hotels and an airport, as most manufacturers have their own planes and fly surgeons in and out of Warsaw. In 1994, the companies employed more than 4,000 people directly, and many others indirectly, for example in metal casting and tool manufacturing.[35] The companies also provide equipment for schools to train students for jobs in the orthopaedic manufacturing industry. Warsaw takes its association with implants seriously, but with a certain amount of humour. At one stage, it boasted a band called Morris Taper and the Reamers; the Morris Taper is a type of hip replacement, and the reamer is the tool for hollowing out the femoral shaft. Warsaw also had a golf tournament called the Lesser Trochanter Open.[36]

The surgeon/company nexus

In the last chapter, and occasionally in this chapter, we have considered the relationships between companies and surgeons. In Britain, these relationships were crucial for the development of the THR. In the US some companies took the initiative in importing and copying models, but through the 1960s and early 1970s many surgeons worked on their own designs and then took them to manufacturers. In his book *Bare Bones*, Augusto Sarmiento wrote: 'At one point, it seemed that any orthopaedic surgeon with a busy joint replacement practice was using his own implant, designed by him in concert with leading manufacturers from around the world.'[37]

Some manufacturers were willing to make any implant that a surgeon wanted, in an effort to improve the quality of implants or develop implants for specific purposes, such as those for fractured neck of femur, or revision cases. During the early stages of hip-replacement surgery, the surgeon's expertise, and that of university departments, augmented a very limited store of knowledge. Through the 1970s, manufacturers hired more scientists and engineers, and their research and development laboratories became more sophisticated, but the surgeon's role was still fundamental. One surgeon remarked: 'when I was involved in the process we were central to the process and we had a lot to do, and nothing was done that...we were not consulted on'.[38]

Yet in some cases, already in the 1970s, manufacturers in the US were controlling aspects of research and development more than was the case in Britain. For instance, well-known Boston surgeon William Harris was prevented by his company from developing his work on new methods of fixation; the company already had a surgeon working on that topic. Harris simply moved to another company and managed to get the new coated stem onto the market before the company that had rejected his research and design ideas.[39]

In 1980 in the US, about nine manufacturers supplied the instruments and implants for much of the orthopaedic industry.[40] Industrial research and development departments increased further in the 1980s, often working with university departments. As a surgeon explained, the companies were 'more targeted towards development while the research at the university (had)... a research component that is understanding the nature of phenomenon'.[41] During the 1990s, American research and development into hip implants was in a large part taken over by the manufacturers, in part because of the responsibilities for design standards then placed on the companies.

The specialisation of implants and the increasing complexity in design also increased the need for project teams as opposed to individuals.[42] New models were given names such as TR28, the Solution, and the Prodigy, rather than being named for a surgeon. The main selling point of Zimmer's Versys is its modularity, and the name is meant to suggest versatility. The specialisation was often reflected in papers given at orthopaedic conferences where some surgeons will focus on the acetabular component and others on the femoral component. As the manufacturing process and the R&D became more complex, the design group was augmented by specialists in metallurgy, chemistry, bioengineering and computer-assisted design. When asked in 2002 about the ways in which the role of the surgeon in the design process had changed, one surgeon said:

> I think it has changed because I think that companies tend to work with teams of surgeons now and I don't think that any one surgeon has had that profound effect on a company as it used to be, the effect that John Charnley had ... the effect that Bill Harris had on Zimmer...no one surgeon can command that kind of control, and

> maybe it's because the technology has improved to the point that nobody's going to come up and say 'Oh, we've got this totally radically new idea', the way John Charnley did.[43]

But individual surgeons continued to be important for the endorsement and marketing of designs, and that continued to raise questions about their relationships with manufacturers. One surgeon said of his role as a surgeon-designer, 'I have never been proprietary. I speak in terms of principles. I mention the implants that I use, but I try to be non-proprietary because I feel that surgeons do not respect you if you are too proprietary. You need professional distance.'[44] Sometimes there is explicit negotiation, for instance when Biomet launched its new Mallory-Head design in 1986, and two designer surgeons were consulted regarding ways to market the implants. Rather like Charnley, they wanted to scientifically test the product with a few surgeons and then get other surgeons to come to them and learn the procedure. The sales manager of that design came up with a plan to accommodate the surgeons and the interests of the company by setting up a workshop at Biomet, and invited surgeons to come and see the system being demonstrated.[45]

But other surgeons act more directly as 'product champions'. They are flown to conduct training sessions on a company's behalf, or to represent the company at a professional meeting. There are different levels of representation: usually only the design surgeons are invited to speak internationally, whilst surgeons who are referred to as 'tertiary' are more likely to speak at merely regional meetings.[46] Most articles presented at the AAOS Congress include acknowledgements of financial support from or research undertaken for companies. The companies have a massive presence at the professional meetings, with elaborate display stands and lavish hospitality; and as we shall see, they also seek influence by substantial funding of training programmes.

Other companies offer clever ways to simplify the management of information. In the *Journal of Bone and Joint Surgery* in 1995, Smith and Nephew offered the Matrix system, a collection of modular hip implants accompanied by a 'Profile Grid' to help the surgeon choose the best implant for an individual patient, and the 'Patient Outcome Measuring System' which allowed the surgeon to 'establish the clinical and economic effectiveness of your implant decision'.[47] The company offered to manage the surgeon's data and customise data-retrieval for presentations.

The manufacturers also work with surgeons over legal and administrative processes for liability and compliance; this bonding is an important side-effect of increased regulation, especially from the 1990s. Dane A. Miller, President of Biomet, reflected on the differences in legislation in the US pre-1976,

> In the US pre-1976 a surgeon could call up and say 'I've got an idea, would you make me a couple of these?' and he'd take them home and put them in his patients. Now no-one in orthopaedics at the time, in my opinion, did just what the surgeon said without applying engineering and material science resources to it, and... I think we would tell a surgeon, 'Hey we don't think that will work for this reason', so it wasn't just a blind make of whatever the surgeon wants without applying some technical input to it. The regulatory systems that exist today, surgeons just can't come in and say 'Hey, I've got a good idea, let me try it', and so to that extent their impact has changed some.[48]

In the early 1990s, the top ten orthopaedic companies increased their regulatory staff by about 40%.[49] Manufacturers conduct all compliance and regulation negotiations with the FDA, on behalf of a surgeon or group of surgeons, for a particular design.[50] If there is a problem with the implant, it is the manufacturer who has to negotiate potential lawsuits, not the surgeon who performed the procedure – unless it is a malpractice case. There is the potential for large lawsuits against manufacturers, and class action against companies.

For formal training too, companies have become more important. While much of the grant money for training orthopaedic surgeons comes from government organisations through the AAOS, manufacturers have also provided funding. In 1969, an advertisement in the *Journal of Bone and Joint Surgery* offered Fellowships to young surgeons to the 1970 AAOS Meeting. To qualify, the young surgeon had to have a letter from the chief of their training programme and a 'scholarly paper of original work'.[51] Zimmer has courted a particularly close association with the AAOS, both in developing a committee to oversee device safety and also in their education programmes. In 1983, Zimmer's parent company, Bristol-Myers, established a grant to the Orthopaedic Research and Education Foundation, which is a part of the AAOS organisation. This educational grant still continues under the Zimmer Orthopaedic Surgery Career Development Award Programme to 'help young orthopaedic surgeons increase their confidence in orthopaedic surgery by advancing their scientific and clinical training'.[52] In-house training of surgeons has now become more common, with companies such as Zimmer setting up their own training institutes, in this case for minimally invasive surgery.[53] As surgery becomes more reliant on technology, the role of the manufacturers in training is likely to increase.

However, the relationship has come under criticism. In 1998, Augusto Sarmiento, then President of the AAOS, in his address to their Annual Congress, argued that surgeons had allowed the industry to dominate. Surgeons had lost power over decisions, and this meant poorer outcomes for patients. Some surgeons have referred to it as a form of 'prostitution'.[54] Orthopaedic surgeons in Britain, such as those at the Nuffield in Oxford, England, have echoed this sentiment, criticising some surgeons for using devices which have not been sufficiently tested, and criticising manufacturers for altering implants while maintaining the same name.[55] Clearly, there is much to be gained for medicine from the expertise and resources of the manufacturing companies, but as these comments show, and as we shall explore in Chapter 6, there can be conflict between commercial demands and the adequacy of clinical testing and control.[56]

In May 2004 in Britain, the all-party House of Commons public accounts committee advised the government to end the system where hip surgeons are 'offered free travel and accommodation by the manufacturers of artificial joints'.[57] The report suggested that one in ten surgeons accepted incentives that had the 'potential to distort clinical judgement'; about the same percentage were using prostheses whose effectiveness had not been proven, and others were doing less than one hip replacement a week.[58] Questions of commerce in surgery thus linked with wider debates about pharmaceuticals.[59]

We will return to these questions in Chapter 6 when we discuss methods of assessment and regulation; but first we look more closely at the ways in which American companies sought to influence and support surgeons, not just through courses, or academic centres, but through the salesmen they made integral to hospital practice.

Representatives and 'Salesman Surgery'

In 1973, at hearings before a Subcommittee on Medical Devices, when changes were proposed to more tightly regulate medical devices in the US, the numbers of salesmen in the field were quoted at 160 for Zimmer, one of the largest companies, and 16 for Sampson, which was one of the smaller companies, located in Pittsburgh.[60] While it is difficult to judge the numbers of sales representatives who now work for medical device manufacturers, as some of them have different types of agreements with their representatives, the numbers of people representing companies who sell medical devices is vast, costing about 20% of the implant price.[61] But the interest lies less in the growth of numbers than in the changing relationships with surgeons.

In his article, 'Attention to "Details": Etiquette and the Pharmaceutical Salesman in Postwar America', medical historian Jeremy Green described the delicately negotiated role of the 'detail man', or pharmaceutical salesman who worked in the hospitals and doctors' offices after WWII, at a time of rapid expansion in the pharmaceutical industry.[62] Marketing was in vogue across the growing consumer industries, and books were published to teach salesmen how to sell. As early as 1941, a new journal appeared, called *Medical Marketing,* aimed specifically at salesmen.

Pharmaceutical sales were largely conducted in doctors' offices, without much reference to individual patients, but some other medical companies became more involved with the day-to-day work of their professional clients, in forms of symbiosis which are characteristic of specialised technologies in the later 20th century. For example, companies like Nuclear Chicago, which provided the new radioactive isotopes and instruments for counting them, also provided technical support in hospitals, and manuals that included accounts of new methods. Companies sponsored meetings for the discussion of new methods; their representatives who visited labs kept abreast of the ways in which the machines were being used, so they could pass the information to other researchers as well as to the company. The reps were central not just to sales and servicing but to a network of information about needs and developments which linked the companies to various kinds of users. In this kind of network, the spheres of commerce are not easily separated from those of academic or clinical research and practice.[63]

With the surgical implantation of manufactured devices, the penetration of commerce into clinical practice became particularly intimate. Many American surgeons learned the use of prostheses, in large part, from company representatives; and that was sometimes true in the UK. Throughout the 1960s and early 1970s, as surgeons developed their techniques, the medical device salesmen were in the operating theatre, advising and also learning. It was in the operating theatre that surgeons could find out if an instrument could be altered to make it more useful, or where a different technique produced a better result. The orthopaedic-implant representative, unlike the pharmaceutical representative, had direct access to that information and could use it to advise other clients as well as the company.[64]

But the relationship could be problematic. In July 1975, two American surgeons, an anaesthesiologist and a nurse were charged with various criminal offences: they had called in a salesman when the surgeons experienced difficulty with a hip implant, and they then allowed him to take over and complete the process.[65] Although the criminal charges were eventually dropped, it came to light that this particular salesman had assisted with several surgical procedures in the Long Island area.[66] In 1977, an unnamed president of a medical supply company told the Medical Practice Task Force of New York State that his salesmen had 'scrubbed in' on 3,240 procedures in five years and had participated in the surgery usually at the surgeon's request.[67] The participation was illegal, but it was not illegal for a salesperson to be present in an operating theatre. The American Medical Association agreed that while a company representative could not touch a patient, they could certainly be on hand, scrubbed up and within the confines of the operating theatre to assist the surgeon with technical questions.[68] In the US, this is seen as a part of the service that the manufacturers can provide to the surgeon, and the presence of the representatives in the operating theatre causes little or no comment.[69] One group of engineers and salespeople estimated that about 90% of hospitals allowed a representative into the operating theatre to provide support to the surgeon.

Nowadays many of the reps are ex-nurses or former operating-room technicians, working for independent distributors who have a contract with an orthopaedic company. The reps may play a major role in planning the operation and the logistics of supply. In general now, the instruments are owned by the distributors and the implants are owned by the manufacturers until they are used; the rep arranges that all the equipment which may be needed will be ready in the theatre.[70] And in as much as they mostly attend complicated or unusual operations, they may be seen as specialists. One marketing manager said,

> These representatives are in surgery every day... That doctor may have one complicated revision a month. The representative may have done three in the last week. And he can say... I saw this happen last week... we need to bring this in... we weren't prepared

> for (this)... and he can really bail the surgeon out of a lot of things.[71]

One surgeon stressed the importance of the representatives in the operating theatre, especially to assist the nurses:

> They do everything but scrub, especially on revisions, and that's the practice in North America now, where the reps go to the O.R. It's not that we need them as much as the nurses need them because everything's becoming so complicated with all this modularity that they're a huge help, and if I have a rep that does not want to come to the O.R.,...I'd phone the company and say, 'Look, I want a new rep'.[72]

But in the 1990s, the role of the salesmen changed somewhat as the American health-care system responded to criticism of huge expenses, especially those borne by employers through insurance schemes for their workers. As Dane Miller of Biomet reported, in the 1980s the focus was more on the technology and the price was less important, but within ten years the focus had shifted more to price.[73]

When hospitals were paid a set rate for hip replacement, it was in their interest to reduce the outlay on the prosthesis. In the mid-1990s, the average implant in the US cost about US$4,000 and the DRG rebate for a hospital was about US$9,800 so the implant was a substantial portion of the Medicare payment.[74] Hospitals seeking to reduce these costs focussed on the supply system, and salesmen found a new role. Hospitals stopped keeping stocks of instruments, which were now supplied by the distributor, and of implants, for which the hospital now paid only when they were used, and not just 'opened'. The hospitals could choose companies that would send representatives into operation rooms so that the particular design and size of prostheses could be decided there, and the companies would bear the cost of the supply system.[75]

As cost constraints tightened, choices about the brands to be used tended to move upwards from the surgeons; exclusive deals were negotiated by hospitals or by a chain of hospitals. Some hospitals formed buying consortia or presented surgeons with directions that in order to receive a bulk purchase price, they would only buy prostheses from one or two manufacturers. One surgeon we interviewed had been told by one of the hospitals where he worked, that he had to choose one supplier for both hips and knees. Because he was so impressed by one company's knee prosthesis, he instructed his hospital to buy the hip implants from that same manufacturer.[76] In this respect, the surgeon is also made into a representative.

Reflections on the business of THR

In the last chapter we saw the Charnley hip taken-up in Europe and the US; it was adapted to new market conditions, and then to a wider range of surgeons and patients. The American conditions encouraged uncemented variants and more expensive metals; and in time, worries about plastic debris would encourage a revival of metal on metal designs. New models proliferated and, as we shall also see in the next chapter, there continued to be a wide range of materials and designs. This was not a system which 'closed' in terms of its technology. But it did close, to some degree in terms of industrial and professional structures. A few firms came to dominate the market, first in the US and then the world.

The firms which succeeded came from different backgrounds. One originated with the production and use of sophisticated alloys, but most were from a succession of orthopaedic companies which had made most of their money from relatively simple products, but had the capital and the connections to develop sophisticated devices. And if even more capital and connections could bring more success, then by the 1980s big pharmaceutical firms were ready to invest. If

4.2 The Madréporique THR: An early French Uncemented Prosthesis made from Titanium

development was easier abroad, and if foreign markets showed local loyalties, then local firms could be acquired.

By the 1990s, most of the big five firms were making the same range of modularised designs. They had become like the pharmaceutical firms to which several of them were connected. They spent lots of money developing me-too products and minor novelties, which then had to be sold hard in a competitive market.[77] Like the pharmaceutical firms, they had huge budgets for marketing and for the securing of 'loyalty' from surgeons. It is hard to get figures for device companies, but it is well-established that pharmaceutical companies spend considerably more on promotion and marketing than on research.

One might imagine that professionals would buy largely on the basis of proven, long-term value, but if that was sometimes the case for some sections of the market, and especially when Medicare repayments were limited, it was not the case for the upper end, where novelty of products and support from the company might be more persuasive. In countries where the provision was more a market than a service, and where even low-risk Medicare patients wanted 'the latest models,' expensive novelties proliferated.

In Britain and the US of the 1970s, the key players were surgeons. But by the 1980s, the companies had taken the lead, and we have seen how intimate were their connections with practice. As in other high-tech areas, companies became major sources of information and major links in the networks between professionals. The companies' representatives were there in the operating theatres, helping the surgeon to chose implants and avoid waste; and they were there in the hospital purchasing and finance departments, negotiating special arrangements for exclusive contracts. A story that begins with instrument companies as adjuncts to surgeons, becomes a tale of surgeons' attendance on companies that increasingly controlled the directions of innovation.

5

Change and Stability in Technical Systems: Materials and Environments

Introduction

In the previous chapters, we have followed the initial development of THRs, especially in Britain, and the development of the hip-joint industry, predominantly in the US. In this chapter we return to the technologies, to examine in more detail how they evolved, and what constrained them. In the next chapter we will relate the technology to its use in health services and to questions of regulation.

First we follow some of the components of THR, and some associated technologies, through to the end of the 20th century and in some cases to the present. We have already noted the recurrent debates about the best materials for the two parts of the prosthesis, and about the best means of fixation; these we now explore in more detail, along with newer issues about resurfacing bone or using biological fixation methods. In the second part of the chapter, we examine Charnley's work on the prevention of infection – for its links with local industries, and especially in relation to the prophylactic use of antibiotics which was popular in the US.

We draw together the debates about materials and designs by thinking of hip prostheses as systems, where the parts interact – so that the choice of material may be linked to choices about design and fixation, etc. In some cases, the choice of one system over another would depend on the patient in question – were they old and frail, or young, heavy and active? It would also depend on the surgical approach preferred by the surgeon in question, which was contingent largely on how he or she had been trained.[1] Since surgical experience, or otherwise, is one of the major determinants of success, this conservatism is more rational than might first appear. But suppose we allow for difference in patients and in surgeons, would we find that surgeons agree on which system works best? Is it clear, for a particular kind of patient and a particular surgical approach, that one system is better than another?

One common story about new technologies is that when they are first introduced, many variants are quickly produced as people search for better models. Then, at least in some cases, selection processes weed out the weaker models, reducing the range to one main type, or perhaps to one type for each part of a now segmented market. The bicycle is a common example: the penny farthing and other variants went to the museums, and new bicycles were thenceforth racers or roadsters; which is not to say that new variants could not appear – mountain bikes for example.[2]

One might expect that when some THRs were shown to work, companies and surgeons would produce a wide range of variants; and indeed we have seen that happen, in the UK and perhaps especially the US. And then, it is true, many of the designs became extinct; but it is also true that many new systems continued to be produced. It was hard to judge which of these were best, because proper judgements required at least five to ten years of data collection.[3] Models that did very badly might be weeded out in a year or two, and there were several such cases; but even then it was sometimes hard to see which of the many variables was responsible for the failure.[4]

Hips could be tested in laboratories, and we have seen that a simulator was developed by John Scales at Stanmore in the early 1970s; simulators are now very elaborate. To some extent THRs could be tested in animals,[5] especially from the 1980s, and the ostrich has proved rather useful. But the workings of the human hip under natural conditions were not well understood, and there was only one ultimate test: how well did prostheses work in the patient, for how long did it continue to work well, and if it began to fail what were the consequences for the patient? Until the 1990s, as the next chapter will explore, statutory regulations did not make much difference, except for the quality of metals and for cements.

If evaluation depends on long-term results, and if modifications of the designs and materials are constantly being suggested, then there are bound to be difficult decisions about how quickly innovations should be clinically introduced. That was true even for Charnley, steadily developing his own models. It was much more the case when many surgeons and companies were keen to introduce their own modifications. Charnley had thought he was designing the equivalent of a Model T Ford – the best design (at least for most of the patients), one that was cheap and where modifications over time would be relatively small. But around a core of old-faithful British designs, as it were, there blossomed a huge variety of systems for which the comparison would seem to be General Motors' annual models and up-market prices. Surgeons hoped for improvements and were attracted to novelties; and especially in the US, cost was not a constraint – indeed there

5.1 Some of the Varieties of THR Acetabular Cups

were good reasons for increasing the cost of the prosthesis, lest it highlight the rest of the charges for the surgery. Companies sold prostheses by introducing new materials and techniques which looked as if they would be of benefit, so the regular launching of new models expected to overcome known problems meant that the range of products remained large. But sometimes the new procedure would bring new problems of its own, and the balance of advantage would depend on assessments of the relative risks.

So the story we tell in the first half of this chapter is only partly about convergence or about linear progress towards a particular mode of THR. For some components and procedures, most surgeons have indeed followed the same path, for example in using cement guns rather than their fingers to pack cement between the prosthesis and the femoral bone, and in using cements containing antibiotics. There has also been a general adoption of modular components to allow more variety and better fitting; and most surgeons would agree that some types of THR are better for particular kinds of patients. But overall, the choice of materials and designs has remained very open. We will follow the changing fortunes of rival metals and plastics, and of different designs – and we will continue the story to cover new 'biological' techniques. We will see that preferences depended in part on the different economic and political systems on the two sides of the Atlantic.

In the second part of the chapter, we will explore the systems around the hip prostheses, and the conditions under which the operations were conducted, especially Charnley's attempts to create 'germ-free' environments in order to reduce the rates of infection. However good the prosthesis and the skills of the surgeon, if the hip became infected, the operation would probably be lost. The patient would survive, but the operation would have to be revised, and it could be very difficult or impossible to get results as good as if the surgery had succeeded initially.

Part One: The material history of THR

The fall and rise of metal on metal

We have seen that McKee adopted a metal on metal articulation for his early hip replacements using a modified Thompson femoral prosthesis with a metallic cup, both made from chrome-cobalt.[6] Initially, a number of well-known surgeons across the globe used the McKee, but they found problems with the design. In order to maintain a smooth head, the metal had to be spherical, which required special manufacturing techniques and these could be very expensive if the prosthesis was made from chrome-cobalt.[7] Whilst Charnley and Craven developed lapping and polishing machines that could improve the sphericity of their stainless steel heads in the hospital workshop, chrome-cobalt was a cast product that was more difficult to machine and required more specialised finishing apparatus. When Charnley released his results for steel and plastic, most surgeons and companies followed suit.

Steel was criticised, however, in the US, where manufacturers preferred the chrome-cobalt already in use for orthopaedic implants. Throughout the 1960s, advertising in both American and British versions of the *Journal of Bone and Joint Surgery* extolled the virtues of Vitallium, using an extensive list of publications to criticise the performance of stainless steel. In Britain, the search for stronger steels proved successful, but some British companies such as Zimmer UK focused on titanium,[8] as did Down Bros. and Mayer and Phelps. Titanium was cheaper than chrome-cobalt, which remained the American favourite.

Even after Charnley's success with polyethylene, some surgeons continued to use or experiment with metal on metal articulation surfaces. We have already discussed Peter Ring in Surrey, who worked with metal on metal until 1979. A modified version of the McKee called the McKee Merle d'Aubigné remained a standard in France, and the Sivash, designed by K.M. Sivash in the late 1950s, was widespread in the USSR, though little used elsewhere. Maurice Müller experimented with a two-component metal on metal articulation made from chrome-cobalt from 1965 to 1968, but then switched to metal on polyethylene as most surgeons were doing at that time.[9] Reasons for the rejection of metal on metal included concerns over potential carcinogenic side effects, possible patient metal sensitivity, the frictional torque issue, high infection rates, and strain on the bone.[10] By the mid-1970s the use of metal on metal was all but abandoned.

In the 1980s however, particularly in Europe, metal on metal began a resurgence. Müller went back to experimenting and designing a metal on metal articulation after he revised six of his early metal on metal operations and found that the prostheses showed very little wear or scratching.[11] While the results of the metal on metal prosthesis were not as good as those of metal on polyethylene, they were certainly not disastrous, and we have seen that the difficulties lay as much with surgical practices as the materials.[12]

Other metal on metal implants followed in Europe from the late 1980s,[13] especially from Sulzer; and in 1995, five different new systems were under evaluation.[14] Their popularity stemmed in part from the identification through improved microscopy of plastic particles worn from the plastic cups in metal on plastic joints. This plastic debris is absorbed by cells and can cause an immune system response, leading to loosening of the artificial joint in a process known as osteolysis.

So surgeons now worried about plastic particles and osteolysis, and it was not obvious how they might be reduced. They also continued to worry about possible carcinogenic effects of metal debris,[15] but metal particles *could* now be reduced – by more refined, and expensive, machining of the metal on metal parts, sometimes using computer-assisted design.[16] Because it seemed impossible to quantify these small and long-term risks, surgeons' opinions were simply added to the complex sets of factors on which decisions between systems were made. Was it worth paying for a much more expensive hip because the risk from plastic debris might be higher than that from the best metal designs, or were these refinements increasing costs for no good reason except that patients and insurance systems would pay for them? Or could one find a material which would be safer than metals or plastics? In the next section, we give a taste of the complex and Gallic history of ceramics in THRs, which will also remind us that the Anglo-American history in most of this book is not the only story.

Ceramic hips

Ceramics were used for hips in continental Europe some years before they were taken up in other parts of the world, partly because the manufacture of ceramics for other clinical applications was more developed in France, Germany and Italy.[17] Patents for use of ceramics in the body appeared as early as 1933 in Germany,[18] and a British patent was taken out in 1961,[19] but it was not until the 1970s that alumina ceramics were used for total hip replacement, by Boutin in France.[20] He had been concerned with the wear properties of polyethylene, so he used an alumina on alumina ceramic for the bearing components. It was first inserted in 1970 with a cemented socket, and a press-fit socket was introduced the following year by Ceraver Osteal of Paris.[21] Even when most surgeons moved in the 1970s from metal on metal to metal on polyethylene, ceramics continued to be used in France, nicely illustrating national differences.

In 1976 the FDA rated ceramic on ceramic articulation as a Class III (high-risk) device which meant that in the US it required investigation and pre-market approval (the only type of ceramic joint approved as a Class II device was a ceramic and polyethylene articulation). The

Autophor, also known as the Mittelmier, was one of the more common alumina ceramic hips in Europe in the 1980s, but if the heads broke and the ceramic fragmented, it could take many surgical procedures to ensure their complete removal;[22] so alumina heads had to be small.

A new ceramic called zirconia was stronger than alumina, but it biodegraded *in vivo* and was also highly radioactive.[23] It was approved for use in the early 1990s,[24] when these problems were circumvented. Many companies now use it in their hip replacements, mainly because it appears to wear less than plastics or metal, it produces less debris, and it provides a smoother bearing surface, which is especially attractive for younger and more active patients.[25] Clinical trials were conducted by Stryker Howmedica Osteonics from 1996 in the US, although the system was already in use in Canada, Europe and Australia; in 2003, it was approved by the FDA – a good example of the new regulatory regime. The Trident Ceramic Acetabular System from Howmedica featured a titanium sleeve;[26] slightly later, Wright Medical Technologies developed their Transcend Articulation Hip System. Yet much of the technology still comes from France, as a 2001 FDA recall of zirconia ceramic heads demonstrated: all of the major manufacturers in the US had bought their zirconia ceramic heads from the same French manufacturer.[27] In this and other respects, the big companies mix and match the products of small specialist companies, some of which they own.

Plastics

Chapter 2 detailed how John Charnley revised the failed Teflon hip cups using acetabular cups fashioned from the new high molecular weight plastic acquired by his engineer, Harry Craven. To recap here, plastics had first been used in surgery of the hip in the 1930s when cups made from Viscaloid in 1933, and then Bakelite in 1938, were used for mould arthroplasties for the Boston surgeon Marius Smith-Petersen.[28] He turned to Vitallium by 1939, but acrylic was used by the Judet brothers in France. The Perspex hip, popularised in the 1940s and early 1950s, was modified by surgeons in Britain with X-ray frames and pins, to strengthen them, and so that any failure could be seen on X-rays.[29] Though the Judet brothers also eventually switched to Vitallium, Lusterlite in Leeds in the 1950s carried a Thompson style prosthesis with an acrylic head.[30]

By the mid-1970s, polyethylene was the mainstay of hip-replacement surgery. Companies selling the polyethylene cup in Charnley or Charnley-Müller designs included Zimmer UK, Wright, Howmedica,

Zimmer, Deloro Surgical, DePuy and OEC. Zimmer also had the Trapezoidal 28 designed by Harlan C. Amstutz in 1969 and Howmedica had the Harris hip. Richards carried some more unusual types: the Tronzo from the early 1970s had a femoral head made of polyethylene, and was also modular, while the St Georg-Bucholz, designed around 1975, had the more standard polyethylene cup.

Originally, the polyethylene cup was a single unit, which worked well, but if a revision was required the surgeon had to remove the plastic acetabular cup from the bone. That could be difficult, and the resultant bone loss could make it difficult to fit a new cup. New double-cup models, or press fit, were adopted in the late 1980s, especially for younger patients who could expect to need a revision.[31] A design by William Harris, for Howmedica in the 1970s, combined a liner made from polyethylene with an outside shell made from metal. The plastic liner was locked in and could be replaced without removing the outer metal shell.[32] The design also meant that the stress on the bone cement was reduced and the polyethylene cup wall could be thicker,[33] but problems emerged in a few cases when the liners became separated from their metal surround and lysis, or bone loss, occurred behind the metal.[34]

The high molecular weight polyethylene introduced in the early 1960s continued to be used extensively in hip-replacement surgery. No better plastic emerged, but by the 1990s, as we noted, research on 'wear' in various combinations of materials appeared to show that debris from polyethylene was causing osteolysis.[35] But the problem, it transpired, was not in fact an old one more minutely described: it was in part new, arising from the sterilisation process. When the polyethylene was irradiated with gamma rays in air, it deteriorated by oxidation; so a new precaution against infection had unexpectedly increased the risk of osteolysis. When this became clear, most firms began to manufacture, sterilise and store their polyethylene in a vacuum, and some took advantage of a different irradiation process which had been shown to strengthen the plastic by cross linking the molecules.[36] One way or another, polyethylene was restored as a material of choice. It was cheap and easily worked. We might say that it had shown Model T qualities, even in the worlds of constant updates and alluring expense.

Cement

Like chrome-cobalt, cement was imported into orthopaedics from dentistry. Polymethylmethacrylate had been developed in 1934 and was used to produce false teeth.[37] Both Haboush and Wiltse had discussed its use in orthopaedic surgery,[38] but cement was first used successfully in Britain, by Charnley, as we have seen. He worked with the local dental school to develop the use of Nu-Life, a cold-curing acrylic denture repair, which he first used for a hip replacement in 1958, as a grout rather than a glue. It was made by a local supplier in Blackpool, CMW Laboratories Limited, which was owned by the surgical instrument manufacturer AD International Limited (ADI). Its high reputation in the bone cement industry and its importance in the Charnley hip system is shown by the fact that CMW was bought by DePuy in 1994 after they took over Thackray – but it still trades as CMW.

While surgeons and manufacturers around the globe embraced cement, much in the way they had polyethylene, in the US the FDA would not release polymethylmethacrylate as anything other than an investigational drug until 1971.[39] So 10,000 surgical procedures with cement were classed as 'experimental' and performed in university hospitals and treatment centres.[40] Charnley welcomed this restriction – not because he thought the cement was dangerous, but because inexperienced surgeons certainly were.

Regulation also shaped the cement business. CMW's cement was distributed in the UK by Down Bros., Mayer and Phelps, and by Deloro Surgical, but CMW refused to provide the FDA with the chemical composition of their cement, which held up its distribution in the US. So the first cement approved by the FDA was the more expensive Simplex cement made by North Hill Plastics in London and distributed (and later bought) by Howmedica.[41] By 1983, when CMW did provide details, Simplex had a strong hold on the American market, along with two American products.[42]

Generally there was little change in cements: 'in the past two decades, almost all aspects of hip surgery have changed but cement has stayed exactly the same'.[43] But that was not quite true. Antibiotics were added and the ways of preparing cement were refined over decades. In the 1970s, the idea of mixing antibiotics with cement was put forward by Bucholz:[44] the antibiotic on the cement bone interface would reduce the likelihood of infection. The resultant product, Palacos R, proved very popular, especially for revision surgery where infection risks were higher. It was also used in two-stage revision operations for failures due to deep infection. In the first stage, the prosthetic components would be removed and the cavity would be packed with the antibiotic-containing cement to help combat the infection. In the second stage, the new prostheses would be inserted after the infection was cleared. Other antibiotic-loaded cements followed, under careful FDA regulation.

There were also incremental changes in the modes of using the cement, which had usually been prepared by hand – a powder and a solution were mixed and the combination was kneaded like dough. Charnley tried various electric mixers and mixing bowls, but he found that the best results were achieved with a spoon and a porcelain pudding basin. Vacuum mixing was introduced by Lidgren in 1984 to protect staff from the chemicals released by the mixing, and because it increased the porosity and tensile strength of the cement process.[45] From the 1980s in the US, cement was centrifuged before it was inserted.[46] In 2003, researchers at the University of Leeds (UK) produced an inexpensive disposable device to replace the thermometer used to measure the consistency of cement;[47] it would inform the surgeon when the cement was at the right consistency to be inserted.

Likewise, there were innovations in the techniques of inserting the cement. Surgeons initially had pressed cement into place with the fingers, but in Boston, William Harris developed a cement gun which was widely taken up in the early 1970s to ensure that cement completely filled the femoral cavity. It was brought to Britain by Robin Ling who designed the Exeter hip.[48] A companion innovation developed in the 1960s was the intra-medullary plug or cement restrictor, placed at the bottom of the prosthesis to stop the cement from progressing too far down the femur. Some UK surgeons developed their own cement restrictors, including Charnley and Ling, but the most successful was a 'shuttlecock' design by Kevin Hardinge which expanded in the femur. Another innovation was the practice of washing out the femoral cavity using water under pressure. The acetabulum, however, was still cleaned by hand and then finger packed.[49] Some surgeons used cement for the femur but avoided its use on the pelvic side, a hybrid combination which was regarded as the benchmark by a panel convened by the (American) National Institutes of Health in 1994.[50]

But for all the advantages of cement and the successive refinements, cement continued to raise problems and some surgeons tried to avoid using it. Reports had appeared in *The British Medical Journal* in the early 1970s detailing a few deaths believed to be caused by the rise in the body temperature when the hot cement was inserted around the prosthesis.[51] The cement was also associated with embolism.[52] By the late 1970s bone cement began to be used less frequently in hip-replacement procedures in the US, and in the early 1980s, the combination of localised osteolysis from cement wear particles, fibrous tissue growth, and the poor strength of cement (which was down to poor cementing technique), caused some surgeons to write about a condition that they called 'cement disease', and which they saw as a major cause of loosening.[53]

Boneloc cement was introduced in Denmark in the late 1980s to try and avoid some of these problems, but it tended to break down, and became a public scandal. The doctor responsible had not done clinical trials, and through to 1995 over 4,000 people were badly affected.[54] Another problem was the barium which for many years was added to cement to make it radiopaque: barium is as hard as stainless steel and particles trapped in the articulating surfaces can scratch the head of the prosthesis.[55]

So while surgeons in the UK continued to use cement in hip-replacement surgery, surgeons in other countries turned their back on another Charnley innovation. Americans were generally suspicious of both plastics and cement. They preferred high-tech alternatives, many of which in fact proved less satisfactory than Charnley's techniques. In the UK, the Charnley and the other cemented models, including the Stanmore and the Exeter, were used in seven-eighths of patients until after 2000. By the 1990s, it was clear that good cement was *not* the major cause of loosening, and that it worked well on most patients.

Coated implants

Whilst cement continued to be popular in the UK, in other countries many surgeons experimented with newer methods of fixation, especially for younger patients. There were two types of cementless implant; one was the press-fit, like the Austin Moore, which relied on the

Table 5.1 Total Hip Replacements per annum, using Cemented or Uncemented Prostheses, NHS in England, 1998–2005[56]

	Cement	Non-cemented	Total	% Cement	% Non-cemented
1998	32,505	4,654	37,159	87.5	12.5
1999	32,975	4,598	37,573	87.8	12.2
2000	32,966	4,858	37,824	87.2	12.8
2001	34,285	5,269	39,554	86.7	13.3
2002	36,355	6,347	42,702	85.1	14.9
2003	38,805	7,926	46,731	83.0	17.0
2004	37,164	10,533	47,697	77.9	22.1
2005	36,329	12,472	48,801	74.4	25.6

mechanical characteristics of the implant-bone interface, the other was porous coatings which relied on biologic fixation to the prosthesis by bone in-growth. Much of the experimental work on the press-fit prosthesis was conducted in continental Europe; that on porous coatings was mainly in the US.[57]

In 1968 Peter Ring developed his cementless prosthesis (in Redhill, UK), arguing that revision was made more difficult by cement and that its long-term effects on the body were still untested. The acetabular cup was screwed into the pelvis and the stem on the femoral side also had fenestrations, which allowed for bone in-growth. The Lord Medreporic[58] prosthesis, named from the rough surface of madrepore coral reefs, was a cementless prosthesis released into the market – in 1975 by Howmedica. The Harris-Galante, manufactured by Zimmer and marketed in the late 1970s, was a forged titanium alloy with titanium porous in-growth surfaces on the stem. There were also in-growth surfaces on the acetabular cup, which had a removable polyethylene liner.[59]

Through the 1980s, Europeans persisted with new press-fit designs: that by Spotorno in 1980 was made from chrome-cobalt and had a spiked proximal part which was replaced by ridges and titanium alloy in 1983. Its design influenced the CLS which was made by Sulzer Medica.[60] Later prostheses included the Freeman, the Mittelmeier and the Morscher and Dick. Like the Ring, threaded acetabular components were part of the design in the Sivash,[61] and the Mittelmeier.[62] There were, however, problems with many of these cementless models, especially on the femoral side; the 'hybrid' – cemented femoral prosthesis and uncemented acetabular cup – became popular in the US, as noted above.

While porous coated implants became more accepted in the 1990s, their roots lay in experimentation throughout the 1970s and 1980s. The first types of porous materials were essentially a type of synthetic bone;[63] other materials such as Ostamer, a resorbable polyurethane, could be polymerised in the body to form a porous structure.[64] None of these materials were suitable for the load bearing properties of bone, however, and other plastic materials were explored.[65] Porous stainless steel was found to corrode,[66] but porous chrome-cobalt alloy and porous titanium worked better. Later combinations from the 1970s included a titanium fibre composite and Proplast which was a combination of Teflon and graphite fibres.

In the 1980s, Ronald Furlong, an orthopaedic surgeon from London had a fortuitous meeting with Johannes Osborn who had used a bioactive ceramic bone material called hydroxyapatite to treat open fractures of the mandible (note yet another connection between orthopaedics and dentistry). Hydroxyapatite could be sprayed onto the implant to improve fixation, and the first Furlong prosthesis coated with hydroxyapatite was implanted in 1985. One of the first cementless designs was the Harris-Galante prosthesis which was marketed in the early 1980s, but there were problems associated with the limited amount of coating on the femoral prosthesis.[67] Studies of these uncemented THRs reported high incidences of femoral bone loss and aseptic loosening,[68] often accompanied by slight and persistent thigh pain.[69]

Thus while many surgeons, particularly in the US welcomed the idea of the cementless prosthesis, especially for the younger patient, most of the early cementless versions enjoyed limited success. New designs which were fully sprayed with plasma, or coated with sintered beads (or later with mesh), improved the results and are now popular for the younger, more active patient.[70] There are very few prostheses on the market in 2006 that do not have some form of roughened surface, if not a coating, to encourage bone ingrowth.

Modularity

Modularity allowed the THR prostheses to be constructed from different components so that the size or material of stem, head and cup could be independently varied according to the needs of the patient. Modularity has been around since the late 1960s, but initially the choices were limited, often to three sizes of complete stem and cup. The Trapezoidal 28, which was introduced by Zimmer in 1969, paved the way to modularity with different stem sizes and neck lengths.[71] The preference for modularity increased throughout the 1970s and 1980s, and by the 1990s, the combinations were almost endless. In 1993, Dow Corning Wright advertised their Infinity Modular Hip System as having 55,000 combinations.[72] Modularity was said to reduce costs as manufacturers did not have to make as many custom models; a stem of titanium alloy might be combined with a head of chrome-cobalt or ceramic, depending on the properties that the surgeon needed for a particular patient.

Bone grafting and blood transfusion

It may be useful to end this section with a note on two rather different materials, where the political economies differ from those discussed so far, but also differ between countries. Blood transfusions and bone grafts are useful, often vital; but they are not normally commodities.

While revision surgery might solve the problem of the implant which had loosened or become infected, severe problems attended repeated revisions; one was poor bone stock or simply a lack of bone for the prosthesis to be attached to. The remedy was to use bone grafts, usually from another human donor, though in some cases from cows. Once the bone graft was accepted by the body it was slowly converted to living bone.[73] Surgeons have performed more complex revision surgery with grafts as the technology has become more entrenched; for example, combining allograft bone with wire mesh to make a sufficient bed for fixing the cup of the prosthesis.[74]

Like many other procedures adapted for orthopaedic surgery, bone transplants have a long history. They were used by 1925, and by 1950 the US Navy had established a Tissue Bank.[75] In Britain, a bone bank was established in 1984 at the Robert Jones and Agnes Hunt Hospital in Oswestry, Shropshire.[76] Between 1993 and 2003 there were more than five million musculoskeletal allografts in the US.[77] This is one area where the manufacturers of the technology we associate with hip replacement do not have any business interests. New kinds of tissue engineering are being developed by companies, but so far, grafting has been a matter for professionals, under national regulation.

One of the reasons for this regulation and specialist control is that for bone, as for other tissues, there were problems of immunity and infection. At the 1973 British Orthopaedic Association Spring Meeting, the subject of bone grafting was raised in association with HL-immunity and problems that may cause in bone grafts;[78] by 2003 potential donors were screened serologically and also physically examined.[79] The first Standards for Tissue Banking were published by the American Association of Tissue Banks (AATB) in 1984,[80] and the FDA has overseen allografts since 1993, although generally the FDA deals with finished products and not the individual technology that go to making up a procedure. By 2002 there were 72 AATB accredited tissue banks in the US which offered demineralised bone products, cortical and cancellous bones and structural bones for hip surgeons who required bones for revision.

The same issues recur for blood transfusion, which impacts especially on hip surgery, which often needs large quantities of transfused blood. There has been particular concern over blood-borne diseases such as hepatitis and malaria, HIV and CJD, and in the US, such worries were aggravated by commercial systems of blood collection, often from populations with low standards of health. The major American solution was to take blood from the patients before the operation – this 'autologous' blood donation has been used by hospitals since the 1970s; patients of almost any age give their own blood once a week for eight weeks prior to their surgery.[81] Where such removals pose dangers, for example in cardio-vascular disease, a further and expensive refinement may now be used, especially in long operations. The intra-operative autotransfusion device allows a patient's blood to be collected from the operative sites, cleaned and returned to the patient. However, the present trend is for strategies to minimise blood loss.

Current developments

Two developments from the 1990s may substantially change THR procedures. Resurfacing of hips is an old aspiration which has become practicable since the 1990s; minimally invasive techniques have been developed in many other fields of surgery and are now being applied to hips.

By another name, hip resurfacing is old technology – as mould arthroplasty as devised by Smith-Petersen in the late 1920s; and we have noted various other attempts to put a cup onto the femoral head rather than removing it, as when Charnley tried with Teflon.[82] However, the technology has been recreated from 1991, and has recently become popular, especially in the US and Canada. In Birmingham, UK, surgeons Derek McMinn and Ronan Treacy founded a small local company, Midland Medical Technologies, to develop a metal on metal, chrome-cobalt prosthesis, with a cemented femoral component and an uncemented, press-fit and coated socket. It was released in 1997, for use in younger patients.[83] It has remained unmodified and in 2005 appeared to be performing well. In 2004 the company was bought by Smith and Nephew for £100 million cash, £33 million of which was for regulatory milestones.[84]

Resurfacing techniques are marketed to surgeons to offer to younger patients, and websites for such patients are now prominent. Patients discuss the pros and cons of hip resurfacing, which is seen more as a lifestyle choice than total hip replacement, which now appears as a procedure for the older patient.

Minimally Invasive Surgery (MIS) for hips was pioneered in the US.[85] Usually two incisions are made, as little as 1.5 inches long; and special retractors are used to push aside the muscles, instead of cutting. In some cases using dedicated technology such as guidance systems, and in others using specialised instruments, the surgeon removes the femoral head in pieces, without dislocating the hip. The new components are then inserted through the other opening. Patients benefit by spending a shorter time in hospital, sometimes returning home the same day, and they have a much smaller scar. The benefit for hospitals is in minimising costs, but there are arguments that the monitoring and care of complication may be more difficult. While MIS appeals to many patients, particularly younger ones worried about the scar, it does require very skilled and highly trained surgeons to work with guidance systems. Medium-term studies, currently being undertaken, are suggesting that MIS is not as effective a procedure as was first hoped.[86]

It is not only surgical techniques that have experienced technological change, and the fascination of the high tech should not blind us to other aspects of the operation and the treatment more generally. Post-operative pain control, without producing confusion in the patient, is important for the resumption of mobility. Anaesthesia can be general, spinal or epidural; short-acting anaesthetics have been used as part of a study at the University of Chicago where 15% of patients were able to bypass the recovery room completely and go directly home.[87] Such techniques may eventually mean that hip prostheses, like new lenses for cataracts, might be done on a day-surgery basis, but of course, much depends on the general condition of the patient. When faced with technical projections, it is always salutary to remember that the subject of the operation is a biological and social being, not a hip, and that much depends on good pre-operative and post-operative care.

Part Two: Surgical environments

Like some other types of surgery which were developed in the 1960s, hip replacement involved long exposures and large wounds, with the additional problem of cutting into bone – the 'headquarters' of the immune system. The operation also involved vigorous activity on the part of the surgeon, and the implanting of a large foreign body – all of which meant that the risk of infection was much greater than in most surgery. The consequences were also more severe: whilst in ordinary surgery infections could be cured by the antibiotics developed from the end of WWII, deep infections of the hip were hard to reach; they caused loosening of the implant and the only solution was for the hip to be reoperated. In effect, the operation was lost; and as we noted, revision operations were difficult and the outcomes less satisfactory. For all these reasons, hip surgery revived issues of infection control which had been topical at the end of the 19th century, but which had

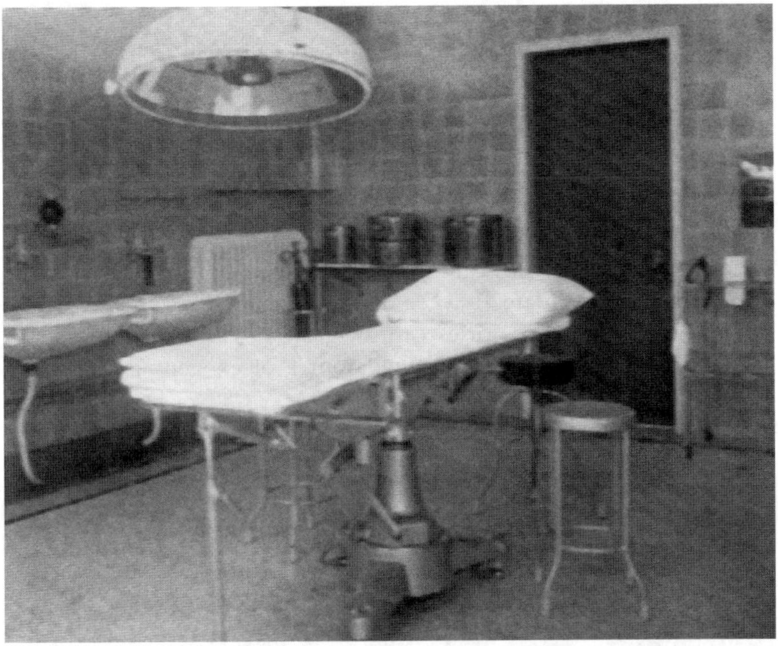

5.2 The Operating Theatre at Wrightington that Charnley inherited in the 1950s

been regarded as overcome by the aseptic techniques developed around 1900 (at least where there was no dirty wound), and for which antibiotics had provided a second line of defence.

In this section, we review the technologies by which the infection problem in THRs has been minimised. We focus especially on the characteristically persistent and systematic work of John Charnley, who returned to the fundamentals of surgical asepsis and used his industrial contacts to develop a germ-free environment. He did so at a time when other hip surgeons, like surgeons more generally, were increasingly relying on antibiotics as a supplement to asepsis. This section thus affords an opportunity to examine two very different, but perhaps complementary, approaches to an intractable problem.

In THRs, infection sometimes became evident two to three years following surgery, and the outlook was grim; in a 1970s study, out of 42 patients with deep infection only eight were able to keep their prostheses.[88] It was sometimes difficult to distinguish infection from 'sterile loosening' attributed to local sensitivity to abrasion-debris produced from cement loosening.[89] The presence of pus-like material that was termed 'sterile infection', was put down to the use of antibiotics.[90] Charnley, however, argued:

> I cannot accept the obvious explanation, in some of these cases, that antibiotics may have sterilised the pus, because under these circumstances the particular antibiotics have not the slightest effect on the healing of the clinical infection in the patient. In other words, we have to postulate that we are dealing with organisms which bacteriological laboratories cannot grow *in vitro*, with some sort of virus, or with an unidentified source of exogenously induced pathology.[91]

That Charnley was sceptical about the power of antibiotics in hip infections, and worried about prophylactic (precautionary) use, was one reason why he sought to reduce the number of germs in the air around the operative site.

Operating theatres built in the 1960s and 1970s were usually designed with plenum air-flow systems, where the operating theatre was at a higher pressure than outside, preventing the inward flow of bacteria-laden air. These systems supplied heated humidified air, but they did not use filters to remove contaminants.[92] In many hospitals, however, including Wrightington, the theatres were older and simpler. In his early cases, Charnley had an infection rate of nearly 9%, which he then set out to reduce.[93]

In 1962, Charnley's technician, Harry Craven, built the first enclosure for use in the operating theatre at Wrightington. Nicknamed 'the greenhouse', it was an operating 'tent' seven feet square and seven feet high, enclosing the patient's body, with air ducted from outside. It was portable, and was set up on the days when Charnley performed surgery. Air flowed in one direction, thus reducing turbulence and the risk of contamination from within the operating space.[94] The prototype for the extraction and ventilation system was designed and donated by Howorth Air Systems, a local firm which usually supplied air filter systems to textile mills or breweries – to remove cotton or nylon dust from the workplace or to reduce contamination of beer vats.

The air was cooled and the current strong, with air changes of 100 times an hour. The enclosure's plate-glass sides allowed observers to see the operation without risking contamination. Only three people

Comparative Figures compiled by the Pathological Department Wrightington Hospital

System	Year	Colonies/ plate/hr.	Total No. of Operations	Infection %
Conventional theatre with exhaust ventilation	1959–1961	80-90	190	8.9
First prototype enclosure Electro-static filter	1962	25	108	3.7
Second prototype Howorth donated system	1963–1965	1.8	1079	2.2
Howorth Permanent Enclosure Mk. 1	1966–1968	0 (limit of accuracy)	1929	1.5
Howorth Permanent Enclosures Mk. I & Mk. II and	1969–1970*	0	2152	0.5
	1971–1972	0	over 2000	0.3
Howorth Demountable System Type DF10	1973–1976	0	over 4000	0.3

Total to end 1976 11,783 operations

* Total Body Exhaust System was introduced only in the second half of 1970, in all enclosures.

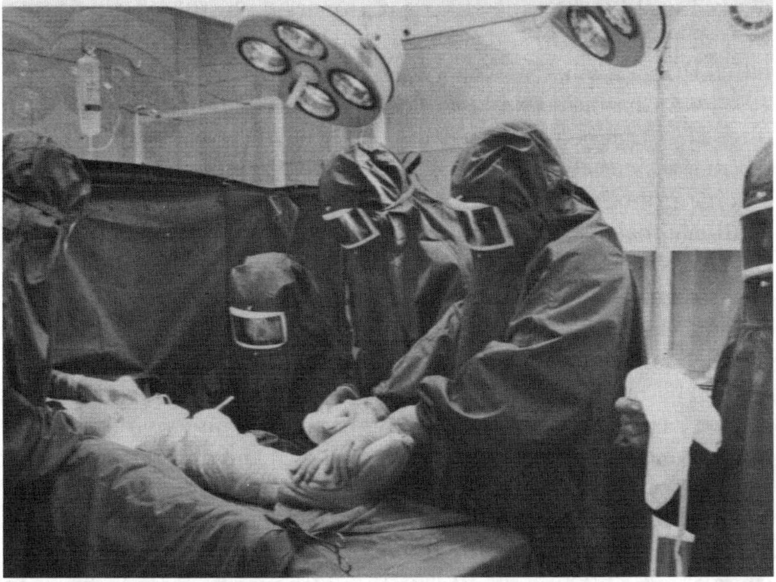

5.3 Howorth Clean Air Theatre Promotional Literature showing the Infection Rates at Wrightington with each Innovation

were allowed in the enclosure (the surgeon, the leg holder and the scrub nurse) and a hatch at one end allowed the trays of instruments to be passed to the surgeon; the patient's head, the anaesthetist and the other theatre assistants remained outside the enclosure.[95] It was used for about two years, and judged to be successful in reducing rates from 6.3% to 1.1%.[96]

Charnley then applied for a grant from the Medical Research Council to build a sterile operating enclosure. He appealed to the Council's desire to be at the forefront of new technological developments, claiming to fear that the UK would be 'overtaken in this field' by others such as Maurice Müller in Berne, who had visited Wrightington and was apparently planning to incorporate a sterile enclosure in his new orthopaedic department.[97] The MRC agreed and a sterile enclosure was installed at a cost of £3,199.[98] This prototype endured many changes before another was permanently installed at Wrightington in 1966. Charnley claimed that the system reduced his infection rate to less than 1%.[99]

By 1973 the enclosure of the Charnley-Howorth air-system, known as the Down-Flow, was ten feet by ten feet; it was manufactured in polycarbonate, which was easier to clean and to move, and which worked well for the expanding export market. By 1976 the side walls were dispensed with, and the new 'Ex-Flow' was marketed as a 'zone' rather than a room.[100] It won a Design Council Award for Excellence in 1977, but Charnley never used the Ex-Flow. He did not like the lack of physical barrier between the patient and the rest of the theatre, and he felt that the curtain of clean air could be easily compromised.

Some orthopaedic surgeons continued to perform surgery in conventional theatres, but a significant number of hospitals purchased the clean-air systems in the 1970s and 1980s. In Britain, Howorth installed 105 systems in the 1980s, including those at the Royal National Orthopaedic Hospital in Stanmore and RAF Ely. A total of 216 were installed in various other countries, the majority in France (47) and the US (38).[101] But Howorth did not have a monopoly on the market for clean air enclosures; some other companies had also seen the opportunities and still others imitated the Howorth and other early designs. In the US, the Laminar Flow Ultra Clean Room System, manufactured by Agnew and Higgins in California and supplied through DePuy, was developed in the early 1970s and priced at approximately US$10,000 (then about £4,000) per unit.[102] Other orthopaedic manufacturers in the US such as Howmedica and Richards also supplied or developed systems.

But Charnley went further, trying to reduce infection from the bodies of the surgeons whose manual tools, such as the brace and bit, required considerable exertion.[103] Research had been conducted as early as 1948 into potential infection by skin particles,[104] and tests in the clean air enclosure at Wrightington demonstrated that the surgeon's skin was indeed an infectious source.[105] So Charnley altered the design of the operating theatre gown, at first adding a hood and then designing a pull-over, or cape style to minimise the possibility of infection through escaping skin flakes.[106]

These early, tightly-woven gowns were unbearably hot. This problem was reduced in 1970 by connecting the surgeon to a hose which removed the heated air and the potentially infectious body exhaust. The pressure in the gown had to be sub-atmospheric so that the bellows action of raising and lowering the arms did not push skin cells into the air. While the body-exhaust suits worked well in preventing infection, the material was still uncomfortable to wear. Woven cotton material was shown to shed more fibres than synthetic fabrics,[107] so new synthetic materials were designed to work like the filters in the ventilation systems. (The same design principles had been used in surgical masks from the 1930s.)[108]

But some surgeons found the body-exhaust suits were uncomfortable and the noise of the extraction system made communication difficult; microphone systems were then installed. Suggested design changes included a gown that tied at the back and a separate hood, but Charnley was unhappy about these proposals, believing that lowering the risk of infection was more important than the surgeon's comfort. He wrote in the *British Medical Journal*, 'What I find alarming is the possibility that surgeons who do not understand the system may put pressure on the manufacturers and purely in the interests of their personal comfort may destroy the efficiency of the installation from the bacteriological point of view.'[109]

In the systematic quest for reduced infection rates, questions were also raised about the lighting. Studies in Sweden and in the US demonstrated that though UV lights did not have a direct effect on rates of infection, they were less likely to heat the air and cause turbulence. Other methods to control infection included 'double-gloving' and different methods of wound closure.[110] Surgical irrigation with saline proved popular in America as a way of flushing out impurities such as blood clots and debris at the operation site. Anti-coagulant drugs were also used extensively, particularly in the US; but Charnley felt they increased infection, so they were abandoned at Wrightington.[111] Other surgeons cast doubt on Charnley's results, suggesting that his lowered infection rates were a result of perfecting his technique, rather than the enclosure and related technologies.

The enclosure, at approximately £8,000 a unit, was relatively expensive to install. The body-exhaust suits were an additional £2,500, which represented a substantial investment for a hospital, and the cash-strapped NHS was often wary about investing in unproven technology. The alternative, antibiotics, were proven to cure most infections, and they did not involve capital investment and hospital space. The risk of bacterial resistance, however, was a concern from the beginning.[112]

Infection which had taken hold deep in the joint was very difficult for antibiotics to clear, but antibiotic prophylaxis was used to curb infection, especially in the US.[113] In one study in the late 1970s, infection rates in hip surgery were reduced from 25.7% to 1.2%.[114] Though it had been demonstrated that laminar flow in joint-replacement surgery significantly reduced the number of infections, a report from the Public Health Laboratory published in the *American Journal of Medicine* in 1981, argued that clean air was no more effective than prophylactic antibiotics or antibiotics introduced into cement.[115] By 1973, there was a 'definite trend' towards the use of antibiotics in hip-replacement surgery in the US.[116] American surgeons seemed to approve of prophylactic antibiotics as a 'just-in-case' measure,[117] and it was later shown that intravenous antibiotics at the time of surgery provided the best form of prophylaxis.[118]

In Britain, antibiotic prophylaxis gained credibility but was adopted more slowly than in the US – mechanical ventilation eclipsing other methods of infection control. Charnley strongly resisted the wholesale use of prophylactic antibiotics until his death in 1982, although he did insert antibiotic powder (0.5 gram each, penicillin and streptomycin) into the cavity of the replacement joint from 1965.[119] In the early 1980s, and in his address to the Hip Society in the US, he advocated the combination of clean air and prophylactic antibiotics.[120]

By 1980 more studies had demonstrated that the results of antibiotic prophylaxis were comparable to the results achieved in a clean air environment.[121] As we have noted, antibiotics, especially gentamycin were included in some cements, but not allowed in the US for many years. French surgeons, like the English, were cautious about antibiotics; by the late 1970s one quarter of surgeons in France were not using antibiotic prophylaxis.[122]

But for all these new methods and refinements, infection continues to be a problem in hip-replacement surgery. The caution of some surgeons over antibiotics has proved justified. Resistant bacteria, such as MRSA, are now a major threat, and clean air continues to be an important part of the response.

Reflections on technical tales

So what do we make of these technical histories? What do they all add up to?

First we note that the dynamics of these technical systems were very heterogeneous. Some elements, such as the cement, remained more or less constant, apart from additives. Some were more or less progressive, in a linear way, such as the improvement of the steels. Some were subject to set-backs which had not been anticipated, and in some cases could not have been, such as the effects of irradiation on the plastic, or a seemingly minor change in surface finish (Exeter) or design (Capital).

It is sometimes supposed that technical systems are better understood and are more predictable than human systems, but of course they have their own uncertainties. Sometimes the 'principles' are clear – but real materials and processes are never just 'principles'; they are idiosyncratic to some degree, the details go beyond the specifications, and the results may be unpredictable. Always in complex systems it is necessary to test, but since the test conditions are more or less remote from real life, it is also necessary to collect feedback, and for long periods, even when the principles seem to be understood.

But for THRs, even the principles were not well-known. Charnley, as an amateur engineer, was convinced that small femoral heads would work better than the 'natural' (anatomical) size; he could not understand why bigger ones seemed to do as well. Some of this ignorance was probably remediable at the time, and some surgeons found engineers and material scientists who knew more than they did about the problems posed. But that was rare – Charnley complained that few academic engineers were interested.[123] Only with the success of the prostheses would biomechanical engineering start to emerge as a special discipline. For the most part, and especially at Norwich and Wrightington, the early work was imaginative craft invention; and as we saw earlier, the patients were sometimes used for experiments, even in the absence of systematic testing of materials. From the 1980s, this kind of clinical experimentation would have been limited by ethics committees, etc. In some cases that might have prevented developments which proved useful; in other cases it would have helped ensure that systematic testing in animals and mechanical simulators removed some of the less likely candidates for clinical experiment.

But part of the problem with THRs was not just the unknowns and the long time periods involved, it was the interactions between new developments and new fears. Debates about the relative merits of metal on metal and metal on plastic have continued for over 40 years, as have debates about the use of cement. This is not just because of different demands for diverse kinds of patients, depending on age and activity; it is also because of changing information, differently evaluated by different groups of surgeons, depending on their histories, skills and national contexts. Compared to the British, the American surgeons placed great store on sophisticated materials, and expense was no object; they were more worried about litigation, and the FDA system interfered with the uptake of cement. The French had a tradition of ceramics to which many surgeons remained loyal.

The swing of opinion according to worries about debris is a classic instance of what we may call spiralling. (It was not a pendulum, from one choice to another and back, because the later technical and social context was not the same as the earlier). Some surgeons worried about metal debris and moved to plastic, but when, with time and new techniques, plastic debris became more evident, some switched back – but to metal on metal models which wore less than the old ones had because they were more accurately prepared. More accurate machining has also helped to establish resurfacing techniques.

But even in conditions of continuing uncertainty, there was some agreement about where the risks lay. For the core clientele of old folk with low levels of activity, Charnley-type prostheses still work well (cemented metal on plastic). For more demanding, younger people, where long-term debris effects are more worrisome and revision is likely, uncemented metal on metal may be more appropriate. But surgeons differ on where they draw the lines, on the range of hips they are comfortable in using, and in their consideration of costs. Under some circumstances, the cheaper models may be more than adequate medically but less appealing economically. Some of the cheaper British hips are not sold in the US, even when they are made by American companies.[124]

Had we restricted our attention to Britain, it would have been a simpler story of Charnley-type designs, refined over time, and satisfying most of the demands – closer to the Model T, and perhaps to most of the other innovation stories about public medicine. What makes the difference in our transatlantic story is the proliferation of innovations relevant for the relatively few British patients who were young and active, and for the many American surgeons and consumers who wanted 'the best'. But when that which is properly known to be 'the best' is *ipso facto* old technology, the best may come to mean 'the latest', and the latest may prove to be expensive failures, as happened with several cementless designs.[125]

This chapter, however, is not just about the devices. One of the main contributing reasons for hip failure is surgical inexperience.[126] That is why surgeons tend to be conservative about their methods, if not always about their devices. Infection is one of the major risks; we have seen how it was reduced, either by generous use of antibiotics, or by elaborate clean-air systems. Again judgements are time-dependent. Charnley did an enormous amount of work to develop clean-air procedures; rather like some of his operative preferences, this could seem rather baroque. It was far easier to use antibiotics, especially since their use for THRs could be justified by the special needs of the operation, and did not make much difference to the general question of antibiotic resistance, even when that was a recognised concern.

It is certainly a concern now, not as a distant threat but as a present danger – so clean-air systems will become more necessary. To combat infection for especially vulnerable operations, Charnley went back to questions about germs that were in some ways late Victorian, and to industrial technologies of a similar vintage. But in matters of infection, as in some others, we are now again, alas, Victorians. Antibiotics no longer shield us as, 50 years ago, we trusted that they would.

6
Hips, Health Services and Quality

Introduction

So far in this book we have discussed surgeons, engineers and companies, and in the next chapter we will discuss patients. Much of this present chapter will be about governments. We will look at the health services of the UK and the US, and the different ways in which THRs existed in those systems. We then turn to discuss issues of quality regulations, from the industrial standards which applied to the materials of hips, through questions of professional assessments, to national registers of hips inserted, or the lack thereof.

Broadly, we are comparing a nationalised health service in the UK with an American system based on markets and on private insurance which does not cover about 40% of the under 65s. In both countries, the technology was subject to little state regulation until the 1980s, because people relied on professional assessments and industrial standards. But from the 1980s, the American pattern changed substantially, in terms of regulation and of insurance; and both aspects revealed the increasing roles of the federal state. Firstly, medical devices were subjected to a form of the pre-release assessment which had been developed for pharmaceuticals. Second was a change in the system of Medicare payments, from 'reasonable costs' to fixed tariffs, as part of wider attempts to contain expenditure. Similar tariff systems are now being used in the English NHS, along with private-sector surgical centres, in attempts to increase competition between hospitals; and the British state is increasingly active in the assessment of technologies – for cost benefits as well as safety.

So, whilst we need to contrast the two national-economies of health, we also need to see their similarities and the convergences, both in services and regulation. For example, in England and Wales in 2004–5 about 34% of THR operations were in private-sector hospitals rather than NHS hospitals, which is not much smaller than the proportion of non-Medicare cases in the US.[1]

Hips and the NHS

Looking back, the Charnley THR prosthesis now seems rather typical of the early NHS – long lasting, relatively inexpensive and reliably manufactured.[2] Once the procedure proved successful, it was thought there would be a backlog of elderly people requiring THR and that demand would then even out through the 1970s.[3] But, as also for most other medical technologies, this did not happen. The procedure was so successful that the patient group increased and demand continued to grow.

The figure below shows a slow increase in total hip replacements carried out in the NHS to the early 1980s, followed by more rapid growth in the 1990s. The gaps in the data and the apparent decline in the mid-1980s are due to a change in the NHS data collection systems. The role of the private hospitals is reviewed below.

From the 1970s, hip replacements in Britain have been a major exemplar of rationing by waiting lists. For the most part, the response was to try to increase the number of competent surgeons and their throughput. Typically, in the 1960s and 1970s, the orthopaedic surgeons had been trained after WWII and they were working in hospitals

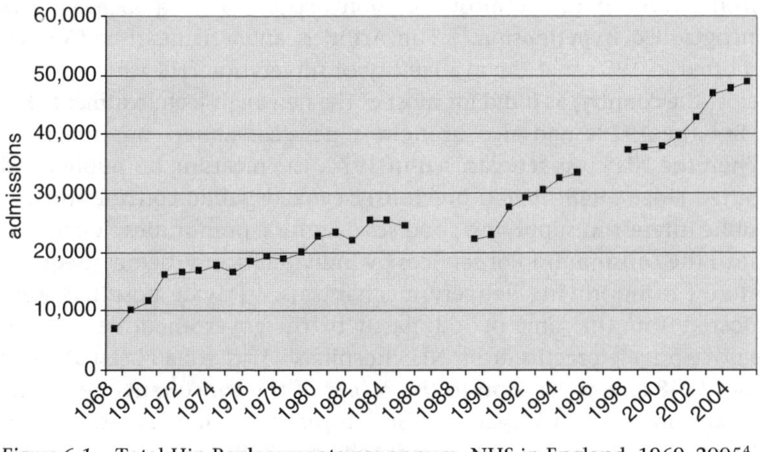

Figure 6.1 Total Hip Replacements per annum, NHS in England, 1969–2005[4]

where there had been no such specialists before the NHS. Surgeons would go to Wrightington to learn the operation with Charnley, or perhaps to one of the other pioneers; or when Charnley's control measures eroded, they might teach themselves his methods, using the fully illustrated book which he published in 1979.[5] Most of Britain's orthopaedic surgeons took up joint replacement, which by the 1970s became a major component of their elective work. The numbers of British orthopaedic surgeons increased by about a third in the 1970s, but membership of the American Academy of Orthopaedic Surgeons increased much more rapidly, doubling from less than 5,000 in 1970 to around 10,000 by the 1980s.[6]

In 1970s Britain, waiting times were greater for hip replacement than for any other operation.[7] Despite a 3% increase in the number of beds devoted to orthopaedics, and a 14% increase in the number of orthopaedic surgeons over the period 1973–78, the numbers of patients waiting more than one year for hip surgery increased to 38%.[8] A survey in 1977 showed a total of 64,806 waiting for elective orthopaedic surgery.[9] In 1981, a report from the DHSS blamed hip replacements for the growth of surgical waiting lists more generally: 'it was the situation in Orthopaedic Surgery which contributed to the deterioration in the capacity of the combined surgical services to keep pace with their waiting lists'.[10] In 1982 *The Times* reported that 20% of patients had to wait a year or more: 'such operations have the longest waiting time of any surgical procedure undertaken by the National Health Service, and patients wait four times longer on average than for other surgery'.[11] Many factors had affected waiting times for hip replacements, including the patients' medical complications; in 1982 the most common reason for postponement was uncontrolled hypertension.[12] The Arthritis and Rheumatism Council reported in 1982 that the availability of surgery for THR varied greatly across the country, as it did for most of the new high-tech treatments.[13]

But the 1970s had also brought a new consumer consciousness. When the NHS was reorganised in 1974, the meaning of 'public' had shifted since 1948. It had previously evoked public control and the public interest in supplying good services for communities; it came to mean the consumption of services by individuals and their consequent interest in improving delivery mechanisms.[14] Private hospitals were boosted from the same period, partly by the government decision to remove private practice from NHS hospitals where it had been allowed from 1948 as a concession to the power of the consultants. As in the US, the increase in private medical insurance came less from individuals than through employment schemes. THRs came to be a major

Table 6.1 Total Hip Replacements per annum, by Source of Funding in England and Wales, NHS and Privately Funded, 1997/8 and 2005

	Privately funded	NHS funded	Total	% private
1997/8	10,707	36,894	47,601	22.5%
2005	13,614	48,267	61,881	22.0%

business for private hospitals; through the 1980s, about a quarter of the THRs were paid for privately. That the surgeons concerned divided their time between the NHS and the private sectors gave them no incentives to reduce NHS waiting lists.

Through the 1970s under Labour, and the 1980s under the Conservative government of Margaret Thatcher, the gross expenditure of the NHS continued to fall behind the medical expenditure of comparable European countries. As international comparison became more common, the NHS ceased to be a model of economy; instead it came to be depicted as a poor system, in both senses of the word. Waiting lists continued to be criticised, but governments were unable or unwilling to invest heavily in staff or buildings. Around 1990, the Thatcher government, which had promoted neoliberal measures across the economy, considered moving to a continental form of insurance, but the NHS remained hugely popular, so instead they introduced various forms of 'internal market'. Primary care doctors could be given discretion over spending on hospital care, and they were sometimes encouraged to seek hospitals with shorter waiting lists.

That the rates of THR operation increased through the 1990s was probably mainly due to more orthopaedic surgeons being appointed, with further specialisation. From c. 1990, in ordinary District General Hospitals, some of the surgeons specialised in lower limb replacements, so throughput may have increased, along with the range of patients thought suitable for operation.[15] Rates have increased again since 2000, due to vigorous attempts to reduce waiting times.

When New Labour took over the government in 1997, it was opposed to markets within the NHS, but by about 2000 this policy had been reversed and neo-liberal principles have been applied with increasing haste, accompanying a major increase in funding. Waiting times have continued to be a major focus of planning and finance, and since c. 2000, as before, THRs have often epitomised the attack on delays. Under European Law, any EU citizen who had to wait for treatment and suffered delays could apply to have their treatment performed in another member state. In 2002, 1,120 people had treatment abroad funded by the NHS.[16] The High Court ruled in 2003 that NHS patients who suffered from 'undue delay' could go overseas and have treatment.[17] Several cases became high profile, for instance when one pensioner took her NHS Trust to the High Court because they tried to make her pay for surgery that she had undergone in France.[18] But orthopaedic surgery still had the largest number of patients waiting over six months; in June 2003, 66,805 patients were waiting longer than six months.[19]

In order to reduce waiting lists for surgery to six months, the New Labour government in 2003 approved the establishment of Special Treatment Centres to concentrate on elective surgery. Freed from the distractions and disruptions of emergency care, they could organise for faster throughputs, as indeed some surgeons had been recommending from the 1990s. Hip-replacement surgery was a key field for which these centres would provide extra capacity.[20] Some of the centres were fully within the NHS, but some were private, often foreign-owned and employing doctors from overseas on short-term contracts.

To get good bulk-deals, many of the private centres drew up exclusive contracts with a single manufacturer to supply all their THRs (e.g. Stryker/Howmedica's contract with Capio, the South African hospital company). In some cases the government forced these Independent Sector Treatment Centres (ISTCs) onto unwilling heath authorities, and the ISTCs were then fully funded for doing relatively little work, with few referrals from local doctors.

The extra capacity, inside and outside the NHS, did indeed reduce waiting times, but the centres also raised worries that by taking only routine cases they were distorting the work and the training in general hospitals. This was especially the case for private-sector centres. Though there was no evidence these performed any better (or worse) than the special centres *within* the NHS, and many reservations about their lack of integration with the main services, a new wave of ISTCs was launched in 2006.[21]

Table 6.2 shows a substantial recent increase in THRs in both independent hospitals and NHS hospitals. The number treated in independent hospitals rose by roughly a factor of three, for NHS hospitals by a factor of two. But if we compare this table with the previous one, we see that this higher proportion is largely due to operations performed in independent hospitals but paid for by the NHS.

The increased capacity brought by special centres and imported surgeons is meant to further increase competition between hospitals – not on charges, for which there are now set tariffs, like American

Table 6.2 Total Hip Replacements per annum, by Type of Hospital in England and Wales, Independent and NHS, 1981–2005[22]

	Independent	NHS	Total	% Independent
1981	6,200	17,464	23,664	26.2
1986	8,091	21,118	29,209	27.7
1997/8	11,226	36,375	47,601	23.6
2004	16,879	30,990	48,987	34.5
2005	21,762	38,189	61,881	35.2

Medicare, but on waiting lists and marginal amenities. In the UK as in the US, the competition between manufacturing companies for contracts with hospitals seems likely to become more intense as the choices of individual surgeons become more restricted and as NHS purchasing authorities take a more direct interest in the cost-benefits of technical choices.

Minimally invasive surgery (MIS) is now supposed to revolutionise joint surgery; the National Audit Office announced in 2003 that the procedure would reduce costs between £15 million and £46.5 million.[23] Most of the economies would come from a reduction in hospital stay but shortening the rehabilitation phase was also expected to reduce the need for nursing care in the patient's home. Problematically, however, most hips inserted in the UK are cemented and the MIS procedure has been designed for the more expensive, uncemented prostheses. Surgeons are still debating whether such minimally invasive techniques can achieve the visibility necessary to fit the components accurately. And of course, little follow-up is available on the long-term efficacy of these procedures.[24]

But we should note in concluding this section and moving to the US, that Britain is not the only country with waiting lists. Some surgeons in Canada have lists of over one year,[25] and the THR delays have become a major political issue: private practice was not allowed to doctors who worked in the Canadian health services, and many young orthopaedists moved to the US.[26] Even in the US, hyper-specialist THR surgeons, such as revision surgeons, can have waiting lists of six months.[27] In Ireland, specialist centres in hospitals have reduced waiting times, but patients can wait up to one year.[28]

THR costs and Medicare

In the US, government concern around hips has been mostly about costs to the Medicare, the publicly funded insurance system introduced in 1965 for people over 65.[29] As those over 65 were still the majority recipients of hip replacements throughout the 1970s, the procedure was generally paid for by the federal government.[30] In 1971, *The New York Times* reported that a hip replacement cost an average of US$5,000.[31] In 1977 a Congressional Committee considered the rates of elective surgery under Medicare, and there was a suggestion that second opinions should be sought to limit the numbers of operations and the spiralling costs.[32] Pressures to limit costs increased in the 1980s, to the extent that they provoked public opposition to 'managed care'.[33]

Prior to the reorganisation of the Medicare reimbursement system in the 1980s, hospitals had calculated the costs of any procedure and simply billed the government for their services.[34] In 1983, the government changed the way that they funded hospital costs, paying a standard lump sum for any given procedure under the new Diagnostic Related Group (DRG) system. Hip arthroplasty was reimbursed under DRG 209, which was still the eighth most common diagnostic group in 1995.[35] The result of this action was a 'sharp drop' in the number of days that patients remained in hospital after their surgery.[36] In one study, the inflation-adjusted hospital cost, between 1981 and 1990, increased only 1.9%, but the hospitals were unable to control the cost of implants, which increased by 117%.[37] By 1997 the direct cost for THR was estimated at US$22,000 per patient, a cost of US$2.5 billion annually.[38] In the same year it was estimated that one-third of all hip replacements in the US were performed on people younger than 65.[39] This was of immediate concern to the private insurance companies, but it also had implications for Medicare provision which would have to cover many of these patients if they later needed revision operations.[40] We do not know the effect of the lack of health insurance on access to THRs from the poorer parts of the US population; but we discuss comparative national rates below.

By the 1990s, orthopaedic surgeons began to accept some responsibility for cost control in American hospitals. In 1992, the AAOS issued a statement entitled *Containing the Cost of Orthopaedic Implants* which suggested that surgeons could work with their hospitals to assist in the control of costs. In some hospitals, surgeons were held responsible for the introduction of new expensive implants, which ensured that a great deal of consideration went into these new systems.[41] We have already seen how some supply costs were passed to the companies.

Surgeons were also pushed to reduce hospital stays. In the 1960s, seven weeks had been the average hospital stay for a hip replacement; though in 1965 McKee and Farrar from Norwich had reported their patients being discharged at two and a half weeks and walking unaided at six weeks.[42] In the 1970s the hospitalisation period had dropped to about five weeks for the average hip replacement. By the late 1990s the hospital stay could be as little as four days, and in a few hospitals, particularly with the younger patient, trials have been conducted with hip-replacement surgery on an outpatient basis.[43] Of course, the time to recover sufficiently to go home varies substantially from patient to patient.[44]

In the early 1990s, because of pressures to contain costs, some manufacturers put forward the idea of a 'generic' hip implant – one example was the Contemporary Hip System manufactured by Howmedica. One study showed satisfactory short-term results with the less expensive implant in elderly and frail patients.[45] In the UK, where more of the recipients of THR are older people, the hip replacements with a long pedigree and history of consistently good performance such as the Exeter, the Stanmore and the Charnley are the more popular hip replacements, and they have low revision rates.[46]

In the US, it is now suggested that low cost implants should be used on low-demand patients. But this is seen as problematic for two reasons that are central to our account: there is insufficient data to determine what is an acceptable level of performance, and because 'Americans would not be happy with the thought that because they are Medicare patients, they would not have access to newer technologies.'[47]

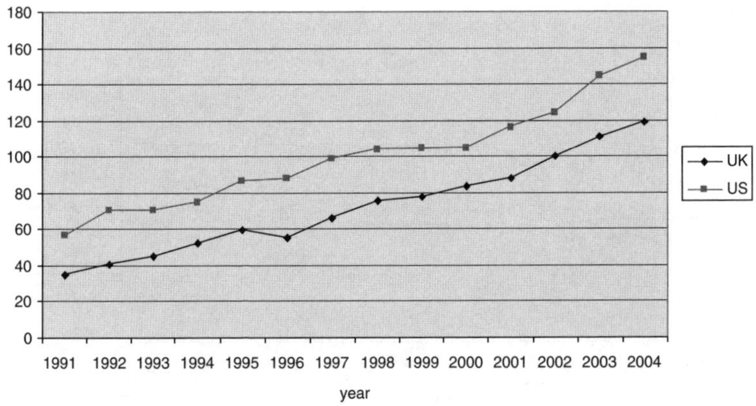

Figure 6.2 Total Knee Replacements (TKRs) per annum, per 100,000 population, in UK and US, 1991–2004[48]

Notes on numbers of THRs in the US and the UK

As these last remarks suggest, and gross expenditure figures would seem to confirm, for most technical medical procedures American rates *per capita* are higher than in the UK. This is true for *knee* replacements where there has been a steady increase in both countries since the early 1990s; the US 'excess' has thereby become a smaller proportion of the British total, from c. 63% in 1991 to 30% in 2004.

But for THR, the graphs are curiously different. The UK, like France and the Nordic countries, now does about 40% more operations than the US (2004) when measured against the total population. The reason lies in part in the age distribution of the populations, and in the ethnic composition. THRs, compared with knee replacements, are mainly for the elderly, and operations on Caucasians are more common than on Afro-Americans; the US population is younger on average and contains higher numbers of ethnic groups. It appears that age correction accounts for about a 20% 'US excess'; the ethnic correction is smaller, leaving some of the difference unexplained.[49] But if we applied the corrections to the 1991 rates, then the corrected US rate would be higher; which would fit the usual pattern of more rapid dissemination in the US. The subsequent more rapid growth in the UK might then be 'catch up' and later extensions of the patient-age range. These are intriguing issues, but ones which await further research.[50]

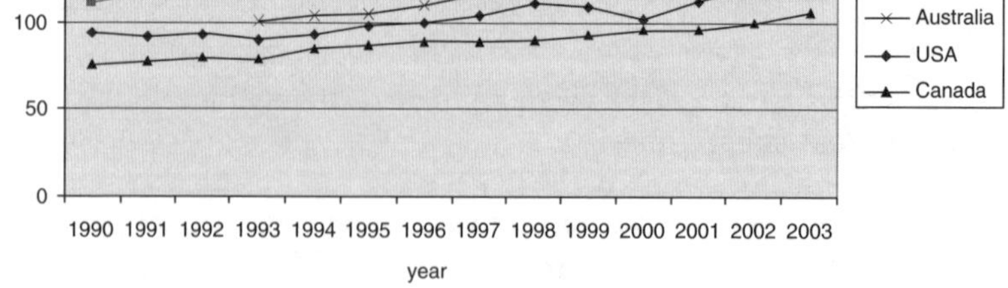

Figure 6.3 Total Hip Replacements per annum, per 100,000 population, Selected OECD Countries, 1990–2003[51]

Costs and benefits

As a percentage of GDP, the US spending on health stood at almost 14% in 1997. This enormous expenditure was peculiar to the US, but in all western countries the exponential rise in medical costs since WWII had attracted attention and concern. (In the UK in 2002 it was 7.7% of GDP but it is expected to rise to 9% by 2008).[52]

From the 1970s, cost-control became increasingly central to health policies in most western countries, especially where insurance arrangements imposed no limit on demand, and where there were plenty of surgeons eager to increase supply. Since that period, in both the US and the UK, calculations have been made about the costs and benefits of medical procedures. Often these were 'back of the envelope', as in mid-1970s calculations about the benefits of genetic screening in reducing the number of babies born with Down's syndrome and Neural Tube Defects who would need long-term care.[53] Such arguments could attract scepticism: a report from the DHSS warned against the application of economies to the cash-strapped NHS in the late 1970s,

> We are of the opinion that too heavy a concentration upon economic considerations when evaluating orthopaedic and other medical procedures must be guarded against... While it may be possible to determine the groups in the population for whom pain and disability might be relieved at least cost, such an approach insensitively applied could lead to the obscuring of the primary purpose of the Health Service and of medical practice which are the relief of human suffering.[54]

But the popularity of such calculations has steadily grown, and the discipline of health economics has become the common discourse of policy makers and planners on both sides of the Atlantic. It developed in the 1970s, in the US, as one of several extensions of business economics into the public sector, and it spread as governments across the West tried to reduce public medical expenditure, partly as a result of the rise in oil prices and shifts in global economics. The main early products were estimates of how many expensive machines, especially CT scanners, would be required for a given population. But, in fact, most services continued to be controlled by demand from doctors, and sometimes from patient lobbies. Calculations might demonstrate that some of these procedures were more effective than others; but that rarely made much difference.[55]

Hip replacement was always an easy case for health economists. THRs are not generally life-saving, but they greatly increase the quality of a patient's remaining years; in technical terms they add to the Quality Adjusted Life Years, or QALYs (a measure first introduced in 1976 by Weinstein and Stason).[56] Of course, it then depends how you calculate the benefits: but if one assumes that a heart bypass will add ten years, and that a hip operation will double the quality of life, over say 20 years, then the benefits will roughly correspond. And if the hip cost is a tenth of the bypass, then the hip will be economically ten times more 'worthwhile'.

The first such estimate for THR, as compared to standard care, appeared in the first UK league table of cost per QALY, compiled in 1985 by one of the pioneer British health economists, Alan Williams.[57] He listed THR with an attractive (low) cost per QALY of £750, based on a revision rate of 2% per year, and it compared well with his other interventions such as £14,000 for a year for renal dialysis and £5,000 for heart transplantation.[58] A US study in 1995, based on modelling and observational data, as opposed to clinical trials, showed that cost per QALY for THR varied by age, with the highest (worst) for the elderly. Even for this group, the estimate of US$27,000 per QALY compared well with other treatments.[59] In Canada at CAN$5,000 per operation, they compared well with the costs of a coronary bypass at CAN$53,000, dialysis CAN$79,000 and transplants CAN$300,000.[60] Much the same was true for Britain in 2000, where an average hip replacement cost £3,686.[61]

Those who liked to work with numbers could also consider the wage-earning capacity or the welfare-cost potential of the patient. As early as 1971, *The Times* reported a study of hip replacement in 27 males; ten had been able to go back to work and their combined income demonstrated that the hip replacement was cost effective; to withhold it was 'crazy economics'. But even in 1990, 74% of patients receiving hip implants in the UK were over 65 years of age,[62] so the economic benefits would be indirect. Several studies suggested that the patient's partners, relatives or friends were freed from the burden of care, and that the demand on health care services, home help and district nursing was lessened.[63]

Given that THR was deemed relatively cost effective, the attention of economists turned to the cost effectiveness of particular aspects of the procedure. Murray *et al.* showed that in 1994, 62 different prostheses were marketed in England with a range of price from £250 to £2,000 – which stimulated interest by the NHS in choice of prosthesis.[64] Most of those used in Britain were at the cheaper end; but the number of uncemented models has risen since c. 2000, and these are generally much more expensive.

The extra cost may be justified if the higher-priced model can be shown to last longer, and to reduce the likely need for a revision. Such assessments demand good long-term data, which is becoming available, especially from national registers, and especially from Sweden which established a register in the late 1970s. But in the UK, a national register has only just been set up, and American orthopaedists still bewail the lack of proper data.[65]

So next we turn to questions of assessment and regulation. Some of the issues have come up in previous chapters where we asked how surgeons measured their successes and failures, how companies dealt with quality issues, and how patients judged the results of their operations. But in this section we focus on governments, especially for the US and the UK, and we end with a case study of a design failure in the UK.

Registers and regulation

When THRs were introduced c. 1960, governments were only marginally involved, but half a century later they were major players, especially in the US. To understand this involvement we have to consider the growth of medical technologies more generally, and the ways in which regulation of medical devices followed from the growing regulation of pharmaceuticals. Hip replacements were not the sole concern of the regulators of medical devices – they were not even a central concern – but the development of hip technology, and hopes for the improvement of results, were bound up with this more general framework. The interaction is of particular interest because of the peculiar characteristics of hip-replacement technology – the lack of good experimental models, the difficulty of predicting the effects of minor changes, the rapid production of new models, and the long time-periods over which the success of the prosthesis had to be measured.

As we have already discussed, the effectiveness of THR prostheses and operations was initially a matter for the surgeon, who depended on the companies for the quality of the prostheses. We have seen that some surgeons and hospitals kept very detailed records and published regularly on their time-series, but many of the new hips were not scrutinised in this way, and even when they were, surgeons were choosing 'advanced' devices that were too new for long-term data. That situation has persisted for 40 years. Only now, in the UK, is national data becoming available on the relative merits of various models over long periods.

Many of the later published articles on THRs were commercially sponsored and their reliability was doubtful. A paper presented at the AAOS Congress in 2003 reported that 47% of the orthopaedic literature surveyed was commercially sponsored, and that good results were reported in 87% of the sponsored articles; but in the otherwise comparable papers without commercial sponsorship, the proportion of those with good results was only 35%.[66]

Thus in 1999 the British Orthopaedic Association could report that the published results offered little to inform surgeons trying to decide which implants to use. The surgeons were influenced by their training, their own perceptions of outcomes, and the service provided by the manufacturers.[67] Increasingly too, they were constrained by the decisions of the hospitals and health authorities, and by budget limitations.[68]

The standardisation of goods

From the beginning, one source of reassurance has been the national standards organisations which operated across many industries in most wealthy countries from the early 20th century. For example, the British Standards Institute (BSI) was founded in 1911 as a voluntary committee drawn from industry, professions, research bodies, purchasers and users.[69] When hip prostheses were developed, there was collaboration between the BSI and the various professional and industrial interests, but not without some jockeying.

In 1955, for example, the Government Department of Scientific and Industrial Research was asked by Zimmer Orthopaedic Ltd. to test a sample prosthesis made from stainless steel as it was felt that the acrylic material was not strong enough.[70] Testing was conducted by J.T. Scales and J. Zarek in the Department of Engineering at Kings College London, who implanted plastic in animals to assess its performance – the work was funded by the Medical Research Council and the Nuffield Foundation. In a letter from the Institute of Orthopaedics at the Royal National Orthopaedic Hospital in London, the surgeon H.J. Seddon wrote:

> Confusion can arise too if a laboratory carries out tests on prostheses evolved more or less empirically by manufacturers of surgical appliances who they themselves have no fundamental knowledge of the problems involved. Of course there is usually a surgeon in the back-

> ground, but that is not a very strong safeguard. Although the conception of these prostheses that have gone wrong was really brilliant, our present troubles are due to a lack of corresponding scientific knowledge.[71]

In later correspondence with the Medical Research Council, Seddon admitted he was worried about the testing because they might have to bring in the Department of Scientific and Industrial Research or the BSI which would mean that medical interests would not be the priority and 'we cannot be ruled by engineers'.[72]

In fact, surgeons and engineers collaborated through the BSI, which set standards for materials and dimensions which were increasingly exacting. For example, the test methods for a cobalt-nickel-chromium-molybdenum material in 1990 included a test for chemical composition, grain size, mechanical properties, ultimate tensile strength, yield strength and elongation.[73] But the BSI paid less attention to design. It provided a quality underpinning, but not a means of deciding which prostheses worked best.[74]

Similarly, in the US in the early 1950s, manufacturers had set up the Orthopaedic Surgical Manufacturers Association (OSMA) to 'promote the patient's best interest' and to standardise basic dimensions and materials for implants and instruments. It joined with the American Society for Testing and Materials (ASTM) in 1964, on Committee F4 which was formed specifically for surgical implants. In the 1970s the committee was headed by Robert T. Rylee II from the Wright Manufacturing Company and comprised 24 manufacturers of orthopaedic implants, instruments and equipment. By 1973, ASTM had published 73 standards relating to dimensional parameters for implants and specifications for materials.[75] But again, it was not meant as a way of judging hip prostheses.

Manufacturers and surgeons seem to have been content with these methods of quality assurance and clinical reporting, but pressures arose from concerns over contiguous technologies – from the intensification of the regulation of medicines following the thalidomide tragedy in the 1960s, and from long-standing worries over medical technologies that were sold direct to patients. We have already seen that regulation of cement in the US dated from the 1970s, because cement was classed as a drug; but effective regulation of the prostheses themselves was not introduced until the 1990s. Because regulation in the UK was generally later and sometimes weaker than that in the US, we begin this section with the American system.[76]

Regulation in the US

In 1997, US$49 billion dollars was spent in the US on more than 65,000 different types of medical devices, manufactured by almost 12,000 different firms.[77] This compares with US$6.6 million dollars, 12,000 medical devices and 1,500 firms in 1972.[78] Before WWII the industry had scarcely existed; medical devices had fallen outside the federal administration created in the early 20th century to oversee 'food and drugs' and especially the vaccines and antitoxins which were so difficult to assess. In the Pure Food and Drugs Act of 1906, there was no mention of medical devices.

It was devices to 'cure deafness' which first attracted the attention of Congress, and under the 1938 Federal Food, Drug and Cosmetic Act, the problem was met by defining the term 'drug' to broadly encompass medical devices.[79] Indeed, the development of FDA regulation for medical devices seems to have been largely driven by 'home use' products sold direct to consumers. The Medical Device Safety Act of 1967 was introduced by President Johnson, worried that without regulation 'the elderly and the seriously ill suffer most. Improper treatment with worthless devices can be the cruellest hoax of all'.[80] But in the 1970s the advanced technologies used by medical professionals also began to attract attention, partly because medical costs had escalated so conspicuously in what the medical industry came to recall as the 'go-go' years.[81]

In 1970 the Study Group on Medical devices, known as the Cooper Committee, issued a report concluding that a new approach must be devised to deal with the myriad of devices that were appearing.[82] Between 1970 and 1976, the FDA established 14 classification panels.[83] In 1973, the US Congress conducted Medical Device Hearings, where they heard evidence from manufacturers and surgeons (the surgeons' meetings were closed). Robert T. Rylee II, president of OSMA and representing the Wright Manufacturing Company of Tennessee reported that if the bill had been passed as law ten years previously, the majority of total hip prostheses would not have been manufactured; if it cost the medical device industry as much as the drug industry to develop an implant then only four companies would be able to do research and development.[84] He stressed:

> All implants are already the subject of the best possible scientific review: namely the continuing observation of a licensed surgeon, who is a far better qualified scientist in these matters than review panels who are not familiar with that patient's particular problems.[85]

He went on to suggest that pre-market approval of devices that had already proved safe was problematic. Of the Austin-Moore femoral prostheses, which had changed little in decades, Rylee claimed 'the only way the prosthesis, *per se*, could hurt you would be to fall off the shelf and hit you on the head'.[86]

In 1976, largely because of problems with a contraceptive shield, the Federal Food, Drug and Cosmetic Act was amended to explicitly include pre-release monitoring of medical devices. The coverage spanned from simple medical items such as band aids (Class I) to much more complicated life saving devices such as artificial hearts (graded Class III, and thereby subject to regulations similar to those for new drugs). The Office of Technology Assessment (OTA) in its 1976 report, *Development of Medical Technology*, looked at the continuous blood flow analyzer, CAT scanners, polio and rubella vaccines, radical mastectomy, anticoagulants, renal dialysis, cardiac pacemakers, cortical implants and the artificial heart. But hip prostheses had been placed in Class II, and no such implants were mentioned in the OTA report.[87]

In 1983 the Subcommittee on Oversight and Investigations published their report *Medical Device Regulation: The FDA's Neglected Child.* It accused the FDA of 'bureaucratic neglect for public health'[88] in only examining Class III technologies. The Medical Device Amendments of 1976 had called for on-site inspections of all medical device manufacturers every two years for Class II devices, but no performance standards had been established for Class II medical devices, including hip replacements. The other major complaint directed at the FDA was that despite promises, it was not communicating with manufacturers when a medical device exhibited an adverse effect. The report concluded that:

> as a result, neither the FDA nor the public knows today how dangerous medical devices are, and we as legislators lack an adequate basis to decide whether the device amendments and their regulatory system are adequate to the task of protecting the public.[89]

The 'amendments' refer to the so-called 510(k) notification under the 1976 Act which allowed 'piggy backing'; a new device did not have to go through all the rigours of testing if substantially equivalent to a device already on the market.[90]

It continued to be the case that hip implants could be modified, under the same name, without the manufacturer having to notify the surgeon of the alterations.[91] Sometimes models with good published results were modified by manufacturers so that a clinically tested design was no longer available. These are not trivial considerations because even minor changes to the design, material, the finish or the fixation techniques can alter the performance of a hip implant, sometimes drastically. The Exeter hip, as we have seen, was compromised when the surface finish was slightly changed;[92] and no one anticipated that the irradiation of high-density polyethylene cups in air would affect the wearing of prosthesis.[93]

Ten years later, in 1994, the American National Institutes of Health Consensus Statement thought it 'imperative to develop standardised methods of assessing and reporting results so that the comparisons can be made between older devices and new "improvements"'.[94] They recommended (in vain) that regional or national registries should be established in order to 'capture a minimum data set on all total hip replacements and total hip replacement revisions'.[95]

Throughout the 1990s, Congress and the FDA were much occupied with legislation and mechanisms to ensure that dangerous failures of medical devices could be recognised quickly, and that the devices concerned could be traced and recalled. Responsibilities were increasingly placed on the manufacturers and the numbers of injunctions and arrests increased notably. But notices about hips were rare. Between 1995 and 1997, the FDA only issued one notice about hip replacements, which was a warning that steam re-sterilisation caused damage to zirconia femoral heads.[96]

The joint industry was, however, more affected by the 1990 requirements for increased pre-market assessment. The FDA had become very cautious after the problem with the Bjork-Shiley heart valve which was manufactured by a division of Pfizer and had been implicated in 300 deaths, mainly in California, in the early 1990s. By 1997 the US General Accounting Office could report device manufacturer's complaints that the FDA was taking far too much time to approve new devices.[97] In the mid-1990s, when the FDA was overloaded with patent applications for medical devices, growth in applications for hip and knee replacements had halved.[98] One venture capitalist told *The New York Times* in 1993, 'Don't screw around with the FDA; let's move these trials to Europe where there's a reasonable process.'[99] What took 13 months for the FDA to approve took 34 days in Germany.[100] Even the Class II devices, which only required a check to ensure that they were sufficiently similar to other products on the market, were taking over 90 days to process.[101]

The Subcommittee of the Committee on Science reported in 1995 that the movement overseas of research and development and clinical trials was alarming because 'the United States has always been heralded as the unparalleled leader in this industry'.[102] It was shown that 46% of manufacturers were increasing their research and development in Europe. Americans would surely be surprised to know Europeans had first access to the newest and most innovative health technologies.[103]

But in the same period, there was also considerable state-sponsored research on hips in the US. In 1993 the NIH was conducting research programmes under the auspices of the National Institute of Arthritis and Musculoskeletal and Skin Diseases (NIAMS). These included patient information, studies of robotic devices in surgery, work on biomaterials (such as the degradation of polyethylene joint components), and on biomechanics such as bone cement interfaces.[104] The Veterans Administration prioritised research on ageing in the 1980s, as the population of veterans from the WWII and Vietnam conflict grew older;[105] and the Agency for Health Care Policy and Research carried out research into the outcomes of hip and knee replacements in order to test the surgery's cost effectiveness and appropriateness for patients.[106]

Regulation in the UK

The *London Evening Standard* reported in 2001 that approximately 2,000 hip joints were recalled following concerns that they were not completely safe, and quoted one surgeon saying that there were at least 20 models used in the UK that he deemed dangerous.[107] Twenty years earlier, in 1981, a paper had been published in *Health Trends* which outlined a surveillance scheme for hip implants that would cover their materials, design, method of manufacture and a limited clinical trial for new models.[108] It was felt that a national joint registry would identify problems early, eliminate certain implants from the market, and track failure rates at specific periods of time.[109] (But even with national registers, the feedback could be slow; for instance 5,000 Christiansen hips were inserted in Sweden before the high failure rate was established.)[110]

Sweden had established a register in 1979; and Norway in 1987, but the only British registers were regional.[111] North-West England, including Wrightington Hospital, had a regional register; another register in the Trent Region covered a population of 4.7 million people, with 66 consultant orthopaedic surgeons. The method of assessment was a questionnaire sent out to the patient after one year, which asked them specific questions about pain, mobility, complications and their overall satisfaction. Of the 80% who responded to these questionnaires,

88% reported themselves as satisfied.[112] Other audits and research studies were conducted in Britain throughout the 1990s, but they were not co-ordinated.[113]

It is not clear just why the 1981 UK register proposal was not adopted. All the relevant parties supported it in public, but they did not press hard and some may have raised objections or dragged their feet in private. It seems likely that some of the major British centres thought their own data was good enough, and that the companies thought a register might slow the rate at which new innovations were introduced. Likewise, some surgeons may have been wary of data that might lead to comparative judgements about professional skills and choices. So they remained reliant on a mix of teaching, personal experience, published data, conference papers, and information from sales representatives. That may have sufficed for some medical judgements; for THRs it was clearly suboptimal.[114]

But since most of the hip replacements in Britain were undertaken through the NHS, and almost all surgeons worked for the NHS, there were other possibilities for data collection and advice which were not present in the US. From the early years of the NHS, there had been a Technical Services Group within the Ministry of Health; it derived from a medical equipment section in the Ministry of Supplies during WWII, set up to encourage the manufacture of medical equipment which could not then be imported. At the end of the war it had inspected and distributed supplies left behind by the US military.[115]

In the 1960s, with the growth of medical instrumentation, product experts were appointed to advise on NHS purchasing, and from 1969 there was a system for reporting defects and adverse incidents in connection with medical devices, plus a Scientific and Technical Board to improve the quality and safety of equipment.[116] In the 1980s, these functions became part of the NHS Procurement Board, which included a Medical Devices Agency (MDA) from 1984. From 1982, the government operated a registration scheme for medical technology companies that complied with the requirements laid down in its *Guides to Good Manufacturing Practice*.[117] This was mainly used to assist purchasers in the NHS; orthopaedic implants were included in the scheme in 1985.[118] The manufacturers had to apply to be registered and to renew their registration every three years.

Regulation, as opposed to purchasing advice, came largely from the European Union, which in 1986 introduced the 'New Approach' for most consumer goods. It was implemented for medical devices in the 1990s, and as in the US, there were three categories, with the most complex and dangerous receiving attention first. Class III, including heart pacemakers were regulated from 1993; for Classes I and II there was a transition period to 1998. But for the most part, the new arrangements did little more than formalise voluntary arrangements which were already in place in the UK. New devices of the simplest kinds were in effect certified by the manufacturers; Class I and II devices were certified by bodies such as the British Standards Institute, or by the manufacturers who were registered with the MDA (which effectively took over the NHS Manufacturers Registration Scheme in 1998).[119]

The MDA oversaw the work of the BSI and the registered companies; it also conducted assessments of certain classes of equipment, in line with NHS priorities. But in contrast to pharmaceuticals, the emphasis was on post-marketing assessment, including reports of faults and adverse reactions. Devices which are registered in any European Union country are accepted in all others; hence the practice of US-based manufacturers developing and registering products in Europe.[120] In 2003, the MDA was merged with the Medicines Control Agency (MCA) to form a larger agency that also covers pharmaceuticals: the Medicines and Healthcare Products Regulatory Agency (MHRA).[121]

Alongside these European arrangements for supervising the acceptance of voluntary standards, the British New Labour government developed a new agency for assessing the cost benefits of pharmaceuticals and devices. The aim was to increase the efficiency of the NHS, rather than prove safety or effectiveness. This National Institute for Clinical Excellence (NICE) issued guidelines on hip prostheses for primary surgery in 2000 and revised them in 2003. In 2000, NICE recommended the use of prostheses that had demonstrated a revision rate of less than 10% in ten years.[122]

The formation of NICE, the success of regional registers, and the pressure for assessment and regulation of medical procedures c. 2000 (including the Capital Hip reports, discussed below), also led to the eventual foundation of a National Joint Registry. The government had initially expected to cover the registry costs with a levy on the manufacturers, who stood to gain significantly from the results; but it backed down when they resisted, and then it refused to fund the scheme from central funds, like all other major national registers. The alternative arrangements, in their complexity and generosity to industry are a nice cameo of New Labour methods. Each private hospital or acute NHS trust pays a levy of £25 per implant to the manufacturers' association, which keeps £2.50 for its own administrative costs, and is invoiced for the rest by the Department of Health, which then pays the fees to AEA Technology, a privatised company which maintains the register from data that is voluntarily provided by orthopaedic consultants. The total annual register cost for hips is about £1 million, but it does seem to work.[123] The first annual report was released in 2004. The number of hospitals participating increased from 138 in April 2003 to 386 by August 2004.[124] Now, in 2006, it covers from 60% to 77% of hip replacement procedures.[125]

A case study: The Capital Hip

We have seen that several THR implants failed at the experimental stage, and that others proved disappointing, for example the cementless versions which caused thigh pain. In Denmark, the Boneloc cement became a national scandal. In the UK, the marker was the Capital Hip; its failure was not dramatic, but it was important, partly in directing attention to the need for assessment and to the difficulties involved.

In the early 1990s, the American company 3M Health Care had produced an inexpensive hip implant for the British market which was sold here between 1991 and 1997. The company, originally Minnesota Mining and Manufacturing, was famous for Scotch (adhesive) Tape but was moving into medical devices. Its design was similar to the Charnley, which had remained the most popular model in Britain, but it provided the surgeon with the flexibility of modularity; it was cemented and the instruments required to insert it were reasonably priced.

But by 1995 the performance of one of the designs was questioned.[126] A paper given at the British Orthopaedic Association noted the higher rate of femoral revision: for one of the Capital designs 5% had failed.[127] 3M Health Care immediately undertook a survey and found the revision rate to be 'within industry standards'.[128] In the Trent Region, 187 Capital Hips were implanted between 1991 and 1995 and at five years had a revision rate of 8.6%, which was considered high when compared with 3% for the Charnley.[129] This was despite the Capital Hip having received the CE mark, which both surgeons and patients accepted as being a satisfactory standard regarding the safety of the prosthesis.[130] Hazard Notice 9801 was issued by the MDA in 1998, and 3M Health Care established a helpline for concerned patients and paid for revision procedures for those patients whose hip implants had failed. They were commended by the Secretary of State for Health in parliament for their 'commitment to provide funding for the assessment, treatment and long-term follow-up of patients.'[131] The Royal College of Surgeons reported that finding patients was made more difficult by the poor quality of hospital records; in 25% of cases, the particular type of Capital Hip had not been recorded.[132]

The problems of the Capital Hip highlighted major complexities in the regulation and testing of medical devices. At this date, clinical results had only been published for about one-third of all current hip designs.[133] In an editorial in the *British Medical Journal*, a surgeon wrote: 'For a new and untested hip to have been introduced into so many clinical centres in such a short period highlights the lack of regulation of both the orthopaedic industry and orthopaedic surgeons.'[134] There were suggestions in the media that the new MDA had been slow to react.[135] The report by the Royal College of Surgeons (RCS), commissioned by the RCS and 3M Health Care, suggested that if data had been systematically collected and analysed, the poor performance may have become apparent by 1993, as was the case for the modular flanged hip.[136]

There have been other THR models withdrawn, for example the Elite Hylamer hip in the late 1990s, when it became evident that wear was excessive and the implants were loosening regularly.[137] The company involved, DePuy, agreed to finance examinations of patients and to pay for revision surgery when an independent panel of surgeons agreed that the hip was unsatisfactory.[138] In 2002, Sulzer Medica, Europe's largest provider of implants, had to pay US$1 billion in a class action by about 2,000 patients whose hip and knee joints were defective due to a manufacturing problem.[139] Some prostheses had been contaminated with an oily residue which prevented the implant from bonding to the bone.[140] Parallel action was taken in countries such as Canada, which established Hip Implants Victim's Aid Associations in Quebec and also launched class actions against the company.[141] The company changed its name to Centerpulse in June 2002 and was taken over by Zimmer to broaden their share of the European market.

Reflections on services and quality

Once again THRs moved between medical worlds. Their importation from Britain to the US involved a significant change of professional, economic and political relations. And as the American situation changed over decades, so too did the British. This was sometimes by deliberate importation, sometimes because the US was 'the future'; its institutions and especially its commercial companies reached worldwide.

When THRs were introduced in Britain, they followed the rules of most medical interventions. The doctor decided what medicines, instruments or implants to use in his practice, and his choices were not the business of anyone else, not even the patient. The manufacturer of medicines or of implants would be legally liable if they were grossly defective, as would the doctor if the dosage or the procedure was demonstrably incompetent; but for the rest the choices were a matter of professional integrity. So careful doctors with new procedures would advance cautiously, using animal experiments where possible, involving healthy volunteers and low doses to check on safety, and trying new procedures on patients for whom there was little hope otherwise. Before the 1980s, it was not necessary to gain the consent of the patient, even for novel treatments. Such was the context in which THRs were introduced, both in the US and the UK.

The main political question in the UK was about the provision of the operation, and the length of the waiting lists, and so it has remained. There has been little debate over the merits of the operation – whether one simply wants to reduce pain and increase mobility, or translates the effects into numbers as increased QALYs or as reduced costs of dependency. For richer old people, this was an operation worth paying for, and the private market expanded. To what extent this market reduced NHS supply, and to what extent waiting lists were accepted as boosting private practice, is a question which awaits further research.

One of the virtues of the new special treatment centres is an equalisation between public patients and paying patients, as waiting lists decrease. It is paradoxical that this attack on the private profits of doctors has been made via the private profits of companies; and it is interesting that the reduction of waiting times seems not to have reduced private demand.

Surgeons knew from the 1970s that the market was filling with designs that were not as well tested as the early models, but neither in Britain nor the US did they do much about it. Some stayed with the tried and tested models, especially for their less 'demanding' patients; others were persuaded of the merits of newer designs, especially for the more demanding. In the Scandinavian countries, registers were launched by the national associations of orthopaedists and paid for by the national government. It seems symptomatic that the British profession failed to follow this example, and that the government intervened around 2000 as the central management of the NHS intensified, and assessment schemes and modernisation plans proliferated.

In the US, the story is one of markets disciplined from the 1980s by tariffs for Medicare payments and by increased pre-marketing regulation. Costs were cut in the 1980s, mainly by reducing hospital stays, and there were some attempts by companies towards supplying more economical prostheses, but the effects on device prices were not large. More generally, the hopes which some Americans placed on tariffs to reduce technical costs have been disappointed. The demands in the 1990s for pre-market testing led to backlogs at the FDA, to the frustration of manufacturers, and to the shift overseas of some development projects.

The European countries continued to rely on professional and industrial standards, but it seems likely that joint replacements will soon be reclassified as requiring pre-market testing. In these matters of regulation, Europe has followed the US; in the provision of registers, the Scandinavian countries led, with the British Commonwealth following, and then Britain.

If we consider all these measures, along with the fact that in both the UK and the US the 1980s saw written and informed consent required for all clinical experimentation and the protocols for trials, then we can see the overall shift: Doctors are to be supervised to ensure that they do not use unknowing patients for research. Their responsibility is shared with other professionals to minimise patient risk and maximise useful outputs. For devices, as for medicines, safety and efficacy are to be demonstrated before widespread use on patients. And if in the US, the federal government often acted by its power to ban interstate commerce, the British government tended to work through the testing, purchasing and reporting mechanisms of the NHS.

7

The Changing Nature of Patients: Expectations and Information

Introduction

> Throughout the day there were more preliminaries. Sister Holmes came and gave me a knowledgeable, motherly talk about the operation. She was followed by one of the nurses I'd seen in the morning, a nice girl who talked to me animatedly. She said something about the prosthesis – 'Haven't you seen one?' she asked. 'Haven't they ever shown you one?' 'No,' I said. I didn't know that I should have been able to look at it, if they had shown me one. 'It looks like a pork chop,' she said...From what I'd imagined, I could see that was what it might well look like. Like a pork chop.[1]

In William Cooper's book *Scenes From Later Life*, the protagonist Joe Lunn has a hip replacement. Joe has more than his share of complications: he suffers a haematoma and an infection and relies heavily on a stick when he leaves hospital, so it is not a particularly successful venture. But his lack of knowledge *is* common: many patients have no idea about what hip replacements look like, and they are poorly informed about the procedure. In general, patients were more aware of the procedure than of the implant. The procedure is felt: the implant as such is not experienced. As long as it is functioning, the patient has little or no awareness or interest in it. But this changed radically if there were complications such as infection or loosening.[2]

This chapter discusses the role of the patient from three aspects; the changing identity of the patients, the developing modes of communication, and the consequences of failures. We try to situate THRs in the changing cultures of medicine on the both sides of the (English-speaking) Atlantic; we analyse how doctor-patient relations have changed since

c. 1960, and how the industrialisation of medical technology has reinforced the consumerism of patients and of the increasingly important media. One way and another, THR patients have become more active!

Changing patients

In the 1960s and early 1970s, most British recipients of hip replacements were 'little blue-haired old ladies'.[3] Initially THR was the preserve of elderly patients who were in pain and debilitated, and whose life expectancy would not surpass that of their artificial joint. At one meeting of the BOA in 1954, a surgeon expressed the opinion that the Judet arthroplasty should be reserved for those 'with a brief expectation of life'.[4] This patient selection protected both patients and the technology by limiting the demands. If a patient was younger and was suitable for hip surgery of any type, they were usually given an arthrodesis particularly if they had a fully functioning hip on the other side.[5] Charnley, in the late 1970s, was typically cautious and distanced about patients under 65:

> The younger the patient the more the surgeon must guard against allowing the patient's subjective symptoms to influence his judgement. The decision to operate should be made almost entirely on the surgeon's objective assessment. He must turn deaf ears to exaggerated adjectives used to describe the intolerable quality of the pain.[6]

That hip replacements were merely for 'old people' was challenged from the late 1980s as younger patients were operated on. And, of course, what constituted old age was also changing as life expectancies increased and people remained fitter for longer, through a combination of better conditions, bodily awareness, new drugs and altered expectations. This wider and more positive view of hips and age was assisted in the US by well-publicised replacements for older celebrities such as Audrey Hepburn and Sammy Davis Jr., and for younger ones such as Arnold Schwarzenegger.

In the US by 1982, 65,000 hip replacements per year were being performed; this rose to 195,000 by 1985 and to about 250,000 by 2003.[7] In 1970s Britain, the only procedure that increased more rapidly than hip replacement was abortion.[8] In Canada there was a substantial increase in hip-replacement procedures between 1981 and 1999, making hip and knee replacements consistently one of the top ten surgical procedures.[9] In the same period there was a 30% increase in patients receiving hip replacements who were under 55,[10] and about a third of American THR patients were under 65.[11] There were 25% more women than men having hip replacements, because of their greater life expectancy.[12] In the US, the THR rate showed significant differences between racial groups, suggesting that osteoarthritis of the hip is essentially a Caucasian disease. The THR ratio for African, Hispanic and Asian Americans is considerably below that for Caucasians, a difference which does not apply to other joint replacements.[13]

Over time the patient base has broadened, but there still have to be strong indications for a patient to have a hip replacement if they are aged below 50. Younger patients commonly include those whose joints have been weakened by excessive sport,[14] and rural workers with strenuous jobs.[15] An article in the *British Medical Journal* in 1997 stated that the younger patient 'overloaded artificial joints with unrestricted activity' and suggested fixing the hip (arthrodesis) as a way to preserve the joint for a hip replacement at a later stage.[16] But the increased sophistication of revision operations, together with efforts to prevent bone loss, mean that a 14-year old with juvenile rheumatoid arthritis may now be given a THR, in the expectation of three or four revisions through their life.[17]

Patient selection may now focus on specific health conditions. One surgeon said that he would not operate on the excessively overweight:

> A 250 pound female is at so much risk because there is so much fat that blood supply to the fat is significantly less, and women tend to put on weight around their hips and thighs, and that is the tissue that is at risk – as opposed to a male who mostly puts it on their stomach. You can have a 300 pound male who is not at much risk from the operation standpoint. I mean they may wear the plastic out – so size in a male (matters long-term) – when you get them over 300 pounds you have to get them on a weight restriction.[18]

In all countries, and for most patients, changes in the surgical management and hospital stay accompanied the popularisation of hip replacement. In the 1960s, patients were rested in bed for weeks before they were allowed to begin the process of rehabilitation. Over the decades the time in bed reduced, so now the patient gets up after one or two days – partly to reduce the risk of deep vein thrombosis and partly because of the beneficial effects of early exercise to strengthen the muscles around the new hip joint.[19] A variety of other factors are also involved.

For another surgeon, one of the biggest factors in the success of THR was the reduction in the number of smokers.[20] By the 1970s the anaesthetics were better and so was the pain management.[21] For example, the cessation of the use of Innovar reduced the incidence of the post-operative mental confusion which could result in non-compliance or a fall.[22] Some companies, as we noted earlier, are now experimenting with MIS – very small incisions and operative techniques which go round the muscles rather than cutting through them.[23]

As recovery times reduced, so expected activity increased. As explained earlier, 1970s patients usually had very low activity levels, but by the 1990s surgeons were recommending that patients could 'intelligently' participate in no-impact or low-impact activities.[24] Hip replacement for younger patients also raised questions about sexual life and childbearing – with care, no problem.[25] But for other patients, the main reason for THR was the desire to play golf.[26] Some patients resumed quite high levels of activity, spurred on by stories featuring Sammy Davis Jr. returning to dancing after his hip replacement,[27] or Bo Jackson, an American football player, who went on to a career in professional baseball,[28] or the Earl of Limerick who climbed the Matterhorn two years after his hip replacement, or the actress Gillian Lynne who returned to the stage after two replacements.[29] The golfer Jack Nicklaus, who featured in many advertisements, had a Stryker ceramic hip named after him in 2003. In one patient guide, cross-country skiing was added to the accepted list of post-operative activities that included walking, swimming, golf, dancing and bowling.[30]

With the increased demands from higher levels of activity on these new hip joints, it became necessary to warn patients against excesses.[31] Hip replacements were not 'as good as new'. As one engineer said:

> I think the Six Million Dollar man on TV didn't help any. You would think the thing would be stronger because it's metal rather than bone. But the bone re-heals itself. The metal just wears down. And you can't do maintenance. You can't get in there and give it a shot of grease.[32]

It is salutatory still to recall Charnley's characteristically Mancunian note of warning in 1961: 'Neither surgeons nor engineers will ever make an artificial hip joint which will last 30 years and ... enable the patient to play football.'[33]

Yet even for less active patients, or those enjoined to be cautious, hip replacement meant regaining independence. One patient wrote to the

British Medical Journal that 'the combined effects of pain and helplessness made me feel profoundly useless, not only to others but also to myself, and relief from this has been one of the most important reliefs of all'.[34]

But younger humans were not the only new patients. Some surgeons tried new hips on dogs, which were also the first recipients of hip replacements implanted by an experimental surgical robotic arm.[35] These were experimental animals, but as early as the 1950s cats and dogs also became subjects of therapeutic interventions. As in humans, replacements of one side of the joint (hemiarthroplasties) met with limited success.[36] Total hip replacements were much more successful, and there is now a huge technical literature on canine THRs. New and expanded veterinary services, such as joint-replacement surgery, increased by almost 8% between 2004 and 2005, taking American pet health costs to over US$18 billion.[37] Artificial hips have also been implanted in more exotic species, including a snow leopard at Calgary Zoo in the 1980s.[38]

Changing assessments

What makes a successful hip replacement? One surgeon suggested that patient compliance was the most important aspect.[39] Another stressed that the relationship is much more than compliance:

> Both the surgeon and the patient must regard the potential surgery as the beginning of a lifetime partnership and in entering into this relationship must exercise the same caution, deliberation and care that one would exercise in entering a major financial partnership or a marriage.[40]

When total hip replacements were introduced, most of the surgeons concerned were already familiar with systems for 'scoring' the success of hip operations. The first grading system for hips had been developed in 1931, and in 1952 it was adopted and adapted by the Judet brothers in France. The results were graded on a scale of 0 to 6 for each of three parameters: for pain, for movement and for walking. Many other systems followed. The British Orthopaedic Association devised one in collaboration with a research group in 1954.[41] The American Orthopedic Association suggested their own rating scale in 1968. Others were developed by Scandinavian orthopaedic surgeons.[42] Charnley's own scoring system for hip replacements, like that developed by the Judets and associates in the 1950s, compared the patient's condition pre- and post-operatively.[43] The Harris Hip Score, developed in 1969 by William Harris from Boston, became and remains one of the most popular assessment tools for hip-replacement surgery. Later alterations were suggested, including the incorporation of separate points for X-rays.[44]

Since the 1990s there has been a move away from accepting the surgeon's assessment of THR outcomes. In newer scoring systems patient satisfaction and quality of life are taken into account and patients themselves complete the questionnaires that are used in measuring outcomes. These questionnaires include a psychometric element that takes subjective elements like pain into consideration. The most popular system in the UK is the Oxford Hip Score.[45]

From the 1970s 'socio-medical indices' became more common: in other words, the effect that arthritis was having on the lives of patients.[46]

7.1 A Consultation – at Wrightington Hospital in 2005

Some doctors used a patient's self-assessment form in part so that the doctor could discuss any particular problems which the patient associated with their condition. And of course, such self-assessment inevitably reflected changes in life-style and expectations. In the 1960s, if a patient had to use a cane to walk after a hip replacement, it would be considered satisfactory; but if a patient in the 1990s depended on a cane and could not enjoy a game of golf, the surgery might be said to have failed. Physicians, now, rarely prescribe a cane or crutches to ease the pain of an arthritic hip;[47] indeed the association of canes with wealth and power throughout the early part of the 20th century has been replaced by the notion of a cane as a sign of dependency and a failed procedure.[48] Patients' expectations continue to increase: one man threatened to sue a manufacturer because his hip lasted only eight years.[49]

Communicating with patients

The bond between a surgeon and a THR patient could be long-term: and for the THR, as for marriage, prospective patients were generally optimistic. The operation gained a good reputation and most patients knew someone else who had benefited. So consultations tended to be brief and one-sided, both with the referring doctors and with the surgeon,[50] and this could be problematic if patients failed to remember or fully understand all that they were told. As in other areas of surgery, there is now a large literature on the problems of gaining 'informed consent'.[51]

From the 1960s, again as with other complicated surgical procedures, hip patients and their families were provided with information booklets, either from the surgeon or the hospital. Larger and more detailed booklets, complete with cartoons and diagrams were provided by the surgeons from the 1980s; but the present fashion, at least in Britain, is for shorter booklets with lots of bullet points – generated by the health authorities, or by charities. In both countries, private companies also provide booklets, with regular updates of the risks involved.[52]

Videos were used from about 1986, usually focussing on post-operative care and exercise.[53] Others from the 1990s used scenarios about 'patients' to demonstrate compliant behaviour; one Wrightington video followed the patient from the initial consultation to rehabilitation.[54]

Television has also added to debates. In *Treat Me Gently* in 1986, a television presenter, Tony Van den Bergh, filmed his fourth hip replacement for ITV. He wanted to remove some of the mystery sur-

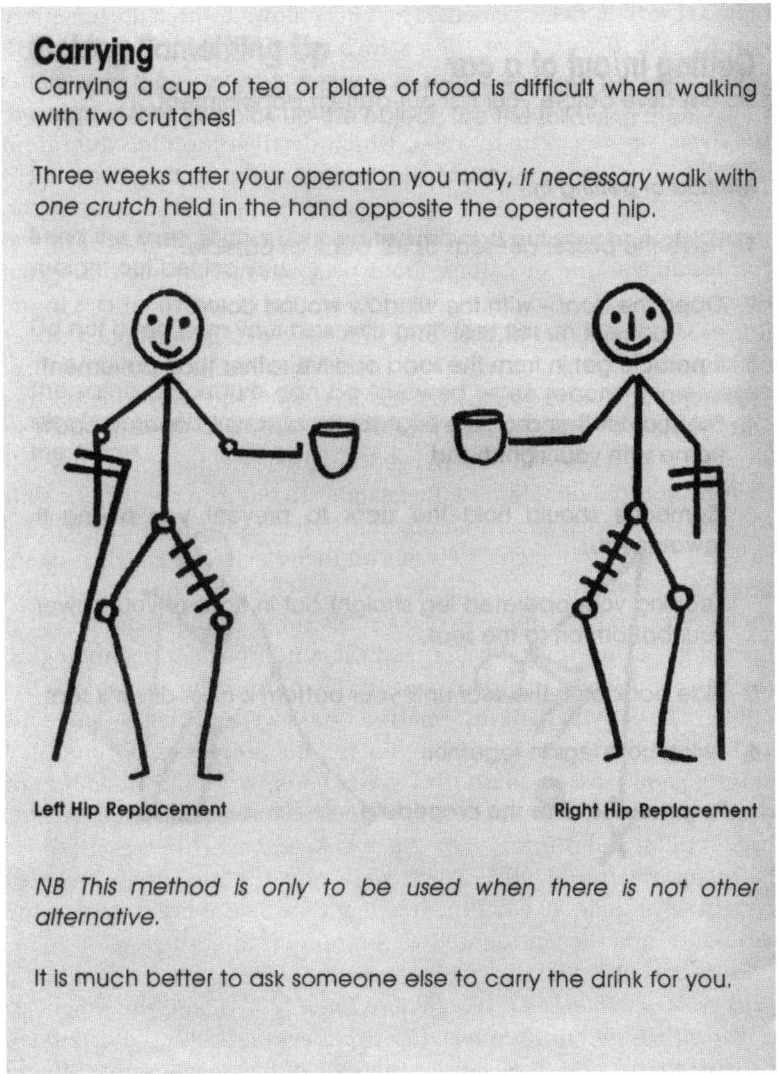

Carrying

Carrying a cup of tea or plate of food is difficult when walking with two crutches!

Three weeks after your operation you may, *if necessary* walk with *one crutch* held in the hand opposite the operated hip.

Left Hip Replacement **Right Hip Replacement**

NB This method is only to be used when there is not other alternative.

It is much better to ask someone else to carry the drink for you.

7.2 A page from a 1970s British Patient Information Booklet 'How to be a Hippie'

rounding surgery, and was praised in the *British Medical Journal* for reducing fear in potential patients.[55] In 1996, ITV's documentary series *The Big Story* included *Celia's False Hope,* which exposed the high failure-rates of the Ring metal-on-metal hip, the Capital hip, and hips replaced with Boneloc cement. The BBC followed with a documentary entitled *The High Cost of Hips*, which also brought failures of THR devices to UK public attention and suggested that hip replacements were not always reliable.[56] American television shows included *Implant: Hip Replacement Surgery* in 1990, which detailed the procedure from initial diagnosis to rehabilitation after surgery.

But even with booklets or videos, patients could not always understand the language used by the surgeon, nor did they always feel comfortable asking questions about personal matters such as sexual function.[57] Such difficulties were nicely explored in two articles in the *British Medical Journal* in 1983 and 1984, one by a general practitioner and one by a surgeon, who had themselves received artificial hips. Their reports gave rise to much discussion, especially from other doctors with implants. Both articles spoke of the unrealistic expectations and anxieties that they had experienced as patients, and which they felt their doctors did not always understand.[58] In our own discussions with American patients, it became obvious that most of them trusted the surgeon to make the right decisions on their behalf, and as the surgery was considered low risk and generally successful, they did not feel that they needed to know more than what the surgeon told them.[59]

For hips, as for several other medical conditions and technologies, the late 1990s saw a major increase in the information available to the public. The early patient information books were written by surgeons and included an overview of hip disease, the procedure, hospital stay and frequently asked questions.[60] Later books written by members of the medical profession included *All About Hip Replacement: A Patient's Guide* (1998) and *Arthritis of the Hip and Knee: the Active Person's Guide to Taking Charge* (1998).[61] These were novel in that they included patient testimonies of what the treatment was like at every stage of the procedure, and they introduced sections on life after surgery.

More recent books for patients include *The Daily Telegraph Hip Replacement: What You Really Need to Know* (2004), and the American *A Patient's Guide to Knee and Hip Replacement* (1999).[62] Written by patients for patients, they reverse the style of the previous texts, giving the patient point of view as augmented by the testimonies of doctors. They also include sections on the alternatives to, and risks of, THR. Both of these books emphasise hip replacement as a choice, and while the older American text focuses on the stages of the procedure and rehabilitation, the more recent UK book also addresses the history and risks in more detail. It underlines its status as a consumer guide by including sections on private treatment and on treatment abroad.

There is also a recent trend to make books even more patient-centred and personal by presenting them in diary style. For example, *Hip Replacement or Hip Resurfacing: A Story of Choices* (2003), recounts a journey through information sources and decisions.[63] Another developing theme is to emphasise the prevention of hip disease and what can be achieved through non-surgical options like exercise – surgery here appears as a last resort.[64]

This reluctance to see THR as the first resort against a painful hip, is underlined by some recent books by US surgeons, such as *What Your Doctor May Not Tell You About Hip and Knee Replacement Surgery* (2004),[65] which emphasise the risks and unknowns of joint replacement surgery and its implants and warn patients against surgeons who insist on hip replacement as a first resort. In a similar negative vein, *My Mother's Hip*, written by anthropologist Luisa Margolies, details her mother's treatment regime and her death five months after a double hip fracture.[66] Thus in the developing market for THR patient-information books, there has clearly been a shift from reassuring, third-person and scientific accounts by medics towards first-person and patient-centred texts which emphasise risks and choices.

British newspapers investigated hip surgery when the late Queen Mother had her hips replaced in 1995 and 1998. After her first surgery, increased numbers of elderly women went to be tested for osteoporosis,[67] and a journalist in the *Glasgow Herald* reported that 'other aged biddies, hitherto nervous and ill-informed, can only gain'.[68] The reports describing the Queen Mother's progress were often accompanied by information about the implant and how it worked, including diagrams. Her second hip replacement, the result of a fall and a hip fracture, demonstrated another major problem of the elderly. That she recovered well from both procedures at ages 95 and 98 respectively, showed hip replacement in a very positive light. But replacement at the ripe old age of 95 was not always an option for NHS patients.[69]

In the late 1990s when formal or informal rating systems appeared in Britain for many government institutions, including the NHS, hip replacement did not escape scrutiny. In *The Sunday Times' Good Hospital Guide*, hospitals were given quality icons as indicators of their suitability for THR surgery. To achieve a symbol of quality, the operating theatre had to have a laminar-flow air system, detailed in Chapter 5, with medical as well as surgical teams at the hospital 'as there is evidence that this improves the outcome for patients.'[70] Hospitals in this same guide also published approximate fees for surgery performed privately, ranging from £5,700 to £13,920.[71] Consumers were becoming better informed about the procedures, if not about the implants themselves.

For hips, as for many other conditions, patients have formed groups, especially since the 1980s, but in this respect hip replacements are low profile compared to many other diseases and technologies. The rise of pressure groups in the late 1960s and 1970s seemed to bypass the elderly who constituted the bulk of implant recipients at the time, but in the early 1980s hip-replacement patients did get together to do walks – demonstrating the hip's ability to transform a patient's life and also raising money for charities such as the Arthritis and Rheumatism Research Council.[72] In 1988, that charity's National Arthritis Week focussed on 'walking miracles',[73] and there was a group called the Prosthesis Club whose members had received replacement joints.[74]

With the rise of patient advocacy and the internet in the 1990s, patients have many more potential sources of information, and the internet can also provide a means for discussion. On chat rooms, such as *Totallyhip* and *Surfacehippy*, patients with hip replacements share their experiences with others and with potential patients.[75] But the participants seem to be younger than the average hip recipients, and many are searching the internet for information specifically on hip resurfacing.[76] Patients can also use other information websites, including those supported by manufacturing companies, or provided by government agencies such as the NHS or NIH. Those more interested in the nuts and bolts of the procedure can log onto the (mainly American) websites which feature live surgical webcasts, and where 70% of the viewers are members of the public.[77] That potential patients may be 'in the theatre', along with the surgeons and the company reps, is a nice cameo of contemporary medicine.

From the middle of the 1990s in the US, the manufacturing companies have used television and then the internet for marketing direct to patients. Advertisements commonly feature celebrities, as we have noted, and they often provide toll-free telephone numbers for a patient to call. They also provide internet addresses, and many manufacturers have websites with specific sections for patients. It is in these interactions that one sees most clearly how hip replacements may become a form of consumption, driven by companies as well as by pain.

Reflections on patients

The British patients who first received successful Charnley hips were old in the 1960s. Most of them had been born in the decades around

1900, in towns where one in five infants could expect to die before the age of five. At the time of their birth, modern surgery was still new and state-welfare measures were just beginning. They ate poor food and breathed polluted air; they were smaller in stature than their children; they aged sooner and died earlier. In the factory towns of the 1950s, there were many new houses, mostly on 'estates' provided by the town councils, but there were also many older houses without running hot water, indoor toilets or electricity. The first generation of THR recipients had borne their children when puerperal fever was still a major risk, but they had lived to see slum clearance, antibiotics, food supplements for children and 'clean air'.

Against that background, it is not surprising that the patients who received Teflon hips from Charnley did not complain that the implants had failed because they had been inadequately tested. They reported instead that they had appreciated the respite from pain and were happy to undergo another operation. Their standard of judgement was their pre-operative condition and any relief from the crippling pain of arthritis was welcomed. They were not 'demanding', as Charnley sometimes complained – when he was not disparaging the excessive demands which American patients seemed to make. Doctors and surgeons were highly respected in the 1950s and 1960s, and British patients seldom questioned their judgements. The standards of these postwar British surgeons were not set by patients, or regulatory authorities, but by professional pride.

By the 1970s, THRs had entered another realm, especially in the US, but also in Britain. It was a world of different expectations, in which patients increasingly thought of themselves as the consumers rather than the beneficiaries of medicine. This culture was older in the US, where public medicine was always in some sense competitive rather than a local service, where national health insurance had been rejected, and where private insurance schemes had grown rapidly between the wars.[78] After WWII, medical specialisation boomed and hospital supply companies marketed new technologies with an assiduity which many other salesmen regarded as exemplary.[79] Patients were referred between would-be specialists, and the hospitals competed both for patients and for doctors by buying far more high-technology equipment than was needed. Costs soared.[80]

That was the American context when THRs were first imported. The cemented versions were at first restricted to academic centres because the cement was covered by FDA regulations. But many surgeons experimented, and many companies were to extend their involvement with prostheses; by the 1980s it was a business. The top orthopaedists could attract elite patients by their own variants of the implants and procedures, and of course the companies were keen to collaborate. In the UK, this emulation was mostly a matter of professional rivalry: in the US it was also, and importantly, a matter of hierarchical markets – who is the 'top man', if you can afford it? If the British system was mostly about increasing the supply for a large pool of older patients within the NHS, the American was about attracting patients and increasing demand. And in both countries, the expectations increased as younger patients were enrolled; they expected more, both because they had more to gain and because they belonged to a less deferential and more demanding generation.

Hence the patient-information model shifts: from what you need to know so that your doctors can relieve your pain, to how you can best hope to restore a level of mobility normal for your age and interests. The level of information shifts – from simply backing-up professional consultations to creating consumers who are informed about opportunities, choices and dangers, and how they can maximise the benefit from the prosthesis.

7.3 An example of innovation in patient information and the promotion of hospital services: The Patient Information Centre at Wrightington, opened in 2006. All images are reproduced by kind permission of The John Charnley Trust.

But hips are not one of the areas where patients came to define themselves by the disability or the procedures they have undergone. The recipients of THRs were not at risk of their lives; they do not, in general, remain on medication; they are not in general 'disabled' in the sense of the disability rights movement. They are happy to have been 'medicalised' because surgery brings the prospect of a near normal life, and if that does not work out in their own particular case, the answer usually seems to lie in more surgery, not in constructing an alternative identity and campaigning for the social world to be made more fitting. The aim of THRs patients is not to live skilfully with a disability or a long-term partial remedy; it is to be cured – but not to push one's luck too far. It is to be ready for relapses – when the further remedy may be problematic, and the more so each time.

8
What is Past is Present

Writing this book has been an adventure in many ways, from the initial proposal conceived with David Cantor on the basis of his research on arthritis, to the work with Francis and the extension into museum projects at the Royal College of Surgeons, London, and at Wrightington Hospital. We did not expect to include as much as we have on patients – that came largely from Julie's interests in disability and cultural history; nor that we would look so much to America, which came from our shared realisation that the UK story only made sense in a transatlantic context. We have ventured into many new areas, and we have lots of questions still outstanding. But in this conclusion we first quickly survey the histories in the UK and the US, and we then draw together some of the lessons we have learned so far.

Developing hips in the UK

Replacements for a *single* part of the hip joint (hemi-arthroplasties) were developed mainly in France and the US around WWII, and they were the basis on which total hip replacements (THRs) were devised and appraised. For many purposes hemi-arthroplasties remain the operation of choice; in 2003 in the US, partial replacements were about half as common as THRs.[1]

But total hip replacements originated in Britain just before WWII, and they were developed here from the 1950s under the NHS. They were part of a general growth of clinical research in postwar, socialised Britain. In some cases the developments were directly linked with strong local traditions in orthopaedics. The most successful of the surgical inventors was John Charnley, a product of the Manchester school, founded by Harry Platt and linked with the US from its beginnings.[2]

Charnley was a science-minded surgeon, and an inventor; he worked with a wide variety of biomedical, engineering and dental colleagues in Manchester University after the war. It was a time of material shortages, but intellectual riches; the war experiences of many of the staff had enhanced organisational skills and opened new possibilities.

The political strength of the University in the new NHS enabled Charnley to create a hospital unit devoted to hip surgery, not in the teaching hospital, but in a sanatorium at Wrightington.[3] It had been built by the County government for the treatment of TB. His work, and that of his able technician Harry Craven, was paid for largely by the British NHS, with some contributions from charity and from the Medical Research Council. He collaborated with a small northern company which made surgical instruments and he continued to draw on a number of local industrial firms and university departments, notably the University of Manchester Dental School. With these resources, and despite set-backs, he created in less than ten years a two-part prosthesis of steel and polyethylene which proved reliable and has been widely used and imitated in all western countries.

The device was not patented, nor was there any government regulation, but Charnley developed the prosthesis and the operation very systematically, collecting excellent data. He arranged courses to instruct other surgeons and he attempted to protect the device by limiting its supply from the manufacturers. He created a global network of surgeons who used his low friction arthroplasty, but that did not prevent a much wider and less careful diffusion of his methods.

His patients were mostly 'little old Lancashire ladies' who were expected thereafter to lead unstrenuous and relatively short lives; they were appreciative and undemanding. At first the patients came to Wrightington; by the 1970s the surgeons trained there were working in district hospitals throughout the region, and more widely. Thus Charnley also created a national and regional centre in a marginal hospital over which he had substantial control.

Charnley's prostheses, along with those from Stanmore and Exeter which owed much to his example, continue in use to the present; indeed they still make up the majority of the British usage.[4] The Charnley, with its improvements and derivatives, remains a standard internationally, partly because it is so well-documented. These THRs were the basis of a major growth in elective orthopaedic surgery. Whilst it would be wrong to present THRs as the sole origin of other joint replacements, they were the first to take-off, and the materials used by Charnley for the THR were used for knees and other joints, notably through the work of the Canadian, Frank K. Gunston, at Wrightington Hospital. It is fair to say that THRs led the way in the transformation of orthopaedics into a specialism dominated by joint replacement and subdivided by joints, as it were. Most orthopaedic surgeons now specialise in hips and knees, or shoulders and elbows, or hands, or backs.

Hip replacements became a model for improving the life of older people by technical means. They were also the paradigm case for waiting lists in the NHS, and they are now central to efforts to reduce waiting lists through special centres for elective surgery. And if THRs have been central to public provision, they have also been a major component of private sector surgery, as that sector has expanded since the 1970s.[5] Patients now expect, not just a reduction in pain and an increase in mobility, but to resume normal life. Again THRs reflect a key feature of modern medicine: that public expectations have risen at least as quickly as medical performance. British patients have become more like consumers, and in that respect among many, they have followed patterns seen first and most strongly in the US.

Developing hips, the American differences

From the early 1960s, Charnley's results persuaded some overseas surgeons that THRs were no longer experimental; they were set to become a standard technique. The Swiss surgeon Maurice Müller immediately moved to develop his own version, with a bigger femoral head and for a different surgical approach. American surgeons and manufacturers were suspicious of Charnley's new materials, but by the end of the 1960s, THRs were being taken up in the US. Many of the product champions were leading orthopaedic surgeons in the medical schools of major cities, and they worked with major American companies which had already invested in orthopaedic technology, or with companies, notably Howmedica, which had special expertise in the medical use of chrome-cobalt (and which also had British links). That Charnley's cement was classed by the FDA as an experimental product until 1971 may have reduced uptake by surgeons not connected with academic centres, and it also encouraged the development of uncemented models.

From the 1970s, the American companies produced many variations and some innovations, especially to limit the use of polyethylene and cement. From the 1980s, THRs were a major business, often linked with pharmaceutical giants, and by the 1990s the business was internationalised. The remaining British THR firms were purchased, but one of the big UK-based medical companies, now Smith and Nephew, moved into joint replacement by buying an American company. In all these global companies, including Smith and Nephew, most of the key orthopaedic staff are American.[6] Some of the main innovations, however, still come from surgeons and small companies, which are then bought up, like the British company responsible for the Birmingham hip resurfacings in the 1990s.

The THR business came to resemble pharmaceuticals in the use of technical representatives to link with surgeons and sell the products. In the US and then the UK, it also followed pharmaceuticals as an object of government regulation; but the technology had characteristics which took it beyond pharmaceuticals in terms of the rapid succession of product-innovations and the promotion of doctor-loyalty. Rapid innovation and surgeon loyalty, however, did not sit easily with the rigorous testing of devices over long periods.

American companies and markets produced lots of new devices and relatively little long-term data. The devices were generally more expensive than the European models, because the cost pressures were often upwards rather than downwards. They were also associated with very profitable markets in surgery, backed by insurance and then by Medicare, which until the 1980s paid 'reasonable costs'. THRs thus epitomised a system where costs rose quickly, and from the 1980s there came the counter-measures: first to restrict payments through Medicare (and through private insurances), and from the 1990s to include THRs with other mid-level medical technologies in more rigorous pre-market testing. The tariff payment systems have helped reduce hospital-stays, but as in other aspects of hospital care, once these savings had been achieved, costs tended to resume their former rise.[7]

The increased pressure to make exclusive deals with hospitals may have reduced some costs, as may supply systems to limit storage etc., but direct-to-patient advertising is set to increase costs as well as demand. None of these new measures seem to provide better long-term assessment, and the collective data in the US remains poor, certainly for hips. American consumers want the latest technologies, even when they are expensive and not well-tested.

What lessons can we learn from this story? We will begin with issues of markets and costs, which will bring us to the question of why British producers lost out. We will then consider the roles of professionals in medical innovation and in the control of quality in health services, which will lead to questions about the role of government. We go on to explore the relevant industry-profession-governments relations in England now. And finally we broaden the view – 'beyond the hip'.

Innovations, medical markets and costs

If we compare the early UK story with the later American, and the surgeon-controlled initiatives with the wider marketing, then one obvious issue concerns expenditures. The work of the British pioneers, in each of the sites, cost a fraction of a single surgeon's salary (the sessions devoted to research rather than service functions), plus a technician. Compared to the cost-benefits of present industrial research and marketing, the economy is evident.

To be sure, the costs which government regulation now imposes on pharmaceutical or device companies are substantial, both for testing and for compliance, and so are the legal liability costs; but for pharmaceuticals, and for implants, much of the development costs go on reproducing products similar to those made by other companies. This intensifies the need to advertise and otherwise promote new products, so the public benefits diminish as a fraction of the expenditure.

The marketing cost for pharmaceuticals in the US are over half the company expenditures;[8] it is harder to get figures for device companies. The costs of 'field-support' are about 20% of the price of a prosthesis, and some of this support would have to be provided otherwise if companies did not provide it.[9] But much of the cost of representatives and marketing arises from proliferation of new designs before they are well-tested and before good comparative data is available to surgeons or hospitals. And because most of the manufacturers, from the 1990s, have made products across most of the range, a surgeon may feel that he can limit himself to the products of one company, likewise a hospital. Thus 'loyalty' is at once more possible, and more remunerative, than for most pharmaceuticals (though similar considerations may apply to specialist internists who prescribe large quantities of a limited range of drugs, for example for diabetes or as immuno-suppressants).

Medical devices have other qualities which accentuate the problem. New pharmaceuticals were relatively easy to protect by patent, and good drugs may be unrivalled for some years; but THRs are relatively easy to copy and to vary. For medical devices generally, new models succeed each other more rapidly than for drugs.[10] Hence the problems with products which were not tested for long enough, or carefully enough. Clearly, suboptimal performance is a dis-benefit to the patients affected, and it will also be a significant public cost if the inadequacies of the prosthesis or operation creates a need for extra care and for revision operations which cost about twice the primary operation.

It seems fairly clear, even to the industrialists involved, that much American development from the later 1970s through the 1980s, was misdirected. This does not simply mean that some of the new models proved problematic, as for example the cementless devices which produced chronic pain; nor does it deny that some of the new prosthetic surfaces proved effective; it means that many new devices were widely introduced in the belief that both polyethylene and cement were bound to be defective, when the clinical data showed that they continued to outperform most of the later variants, at least for older patients. And that by the end of the 1990s several manufacturers were reintroducing older, cemented models.[11]

The engineers liked elegance, and industrialists in these markets liked novelty; looking back, we can see that the novelty was overdone. We can also see that this could have been avoided, and that the state medical systems of the UK and the Nordic countries reduced both the hazard and the counterproductive expense. Even now, costs in the US could be cut by going back to older devices, which are more than adequate for the majority of recipients.[12]

Markets and nations: Did Britain lose out?

That the successful companies were mostly American, and that the British companies were eventually bought up by multi-nationals, raises questions familiar from other studies of new technologies, including such medical technologies as the intra-ocular lens (IOL) replacements for cataracts, or computed X-ray tomography (CT scans). Both are British products which, like the THR, took off in the US from the early 1970s.

For intra-ocular lenses, the breakthrough was at Moorfields Hospital, London, where in 1949, Harold Ridley first inserted a plastic (Perspex) lens into the sac from which the clouded lens had been removed. But the technology remained problematic until the late 1960s when an American surgeon-inventor found a method for liquefying the old lens so it could easily be removed, and then for inserting foldable lenses. Ridley had worked with Rayner, a small ophthalmic company in London, and with Imperial Chemical Industries which supplied the Perspex.

In the late 1960s the intra-ocular lens proved controversial in the US, partly because the data was so poor; it became an important example for the regulation of devices by the FDA[13] and Rayner benefited from theirs being the first such product so endorsed (1981). They continued to expand in Europe, but they rapidly lost out in the US market during the 1980s. In the 1990s, as for THRs, most of the small IOL companies were bought up by multi-nationals based on pharmaceuticals. But unlike Thackray and the other UK manufacturers of THRs, Rayner remained independent and in 2003 re-entered the US market.[14] Here, as for the recent Birmingham (UK) hip-resurfacing technology, we are reminded that the UK surgeons and companies continue to be creative and that, contrary to much of the business rhetoric, successful product innovation does not require the huge resources of the multinationals – whose advantages lie more in the marketing.

In his study of CT scanning, the historian of technology, Charles Susskind, described how the British music company EMI, rich with profits from the Beatles, invested in the ideas of one of their engineers, Godfrey Hounsfield, and developed X-ray computer tomography as a medical technique.[15] They worked closely with academic medical centres and various government agencies. Technologically it was brilliant, and it rapidly became attractive in most rich countries. Indeed, as Stuart Blume emphasised, it became a classic case of 'oversupply', when social scientists came to analyse the usage of the scanners. But EMI soon lost out in America. Attempts to control excessive expenditure depressed the market, and in the mid-1970s established equipment companies with good arrangements for marketing to hospitals poured huge capital into rival products. EMI, and several other companies, including a major pharmaceutical firm, were forced to withdraw. The CT business of EMI was bought by GE, the American electrical giant which, together with Technicare (a part of Johnson & Johnson) then dominated the US market in CT technology.[16]

Compared with CT scanning, it is predictable that British THRs did not fare well in the US. Charnley's devices were easy to imitate and modify, and there was a strong tradition in both countries of surgeons promoting their own instruments and devices. To the extent that THR was derived from the previous single-component hip replacements, several American companies already had the technologies and the connections to move into THR production and marketing. Some of these companies were indeed established in the UK market before THRs took off, and thus were in a good position both to import British knowledge to the US, and to export American products to the UK. If we note that the US market was six times that of Britain by population, and that the medical expenditure *per capita* was considerably higher, and add that from the 1990s the FDA regulations encouraged American investment in Europe, then the loss of British firms is anything but surprising.

There the analysis may end if we are doing economic history; but because of its direct link with individual and public welfare, history of *medicine* calls for other assessments. The conditions of medical technology markets determined the success of the big American companies, but that does not mean that the public gained thereby. We have shown that markets and industrial quality control, together with the normal discretion of professional users, proved insufficient for the assessment and control of devices and procedures which were very sensitive to small changes and which require long-term evaluation. The main additional controls were twofold: professional organisation and the agencies of government. We will discuss them in turn.

Innovations and professions

Many authors who have written on medical innovations have noted the importance of professional networks as both promoting and controlling the spread of novelties. In general, doctors follow people they trust. They know that Dr X has a reputation for picking up on new things, but they will not generally be persuaded until Dr Y supports the case – Dr Y having a position of authority, probably in a medical school. If this is true for easily-used medicines, it seems to be truer for new surgical devices and operations. These can perhaps be learned from books, and we have heard of surgeons who performed Charnley operations with the book propped open in the theatre; but surgeons will usually try to see an operation performed before they try it themselves. The more conscientious, especially if they are going to train others, will pay extended visits to learn major new techniques, as Charnley encouraged at Wrightington. Thus the spread of a procedure relates to the hierarchies of medicine, to master-pupil relations, and to the links between regional medical centres and their peripheral hospitals. It is not simply a matter of markets.[17]

When American companies took up THRs, they did so in conjunction with major figures in the American profession. Several leading American surgeons worked with Howmedica, the firm which specialised in chrome-cobalt (Harris and Turner in Boston; Engh in Baltimore). The Richards company collaborated with Bechtol in Los Angeles, whilst the Zimmer company worked with Galante in Chicago and then with Harris in Boston. The DePuy company's initial venture was with Müller in Switzerland for whom they were the American agents, and they later worked with Engh.[18] Initially, the companies were highly dependent on these centres – for clinical input and for the prestige they supplied. But as THRs became established, marketing increased in most countries, and professional control became more difficult, especially if surgeons saw themselves more as commercial rivals than as partners in an extended programme to develop a technology. This was especially the case in the US where commercial pressures were higher and academic hierarchies lower. Though companies remained linked to centres of excellence, they depended on them less and came to treat them more as sites for trials than as centres of research.

It is interesting to compare THRs with the direct fracture treatments developed by the Arbeitsgemeinschaft für Osteosynthesefragen (AO), which was directed by a group of Swiss surgeons who franchised and closely controlled the manufacture and sales of the instruments, and the metal plates and screws, etc. In his history of the company, Thomas Schlich claims that American surgeons resisted the new methods because they were not ready to buy into the AO ideology, which subordinated individual surgeons to the standards of the whole enterprise – requiring, for example, that they passed a test of competence, and that they submitted records of their work. Swiss surgeons would conform because they recognised AO as a surgical fraternity, and West German surgeons because they respected academic authority and were organised, at least informally, as graduates of major centres, into which AO was accepted. The East Germans accepted AO standards when these were adopted by their national authorities; and quality control was well-maintained there because in conditions of scarcity only the well-qualified received the kit. But in America, none of these structural forces were as strong, so most American surgeons worked as individuals, who judged for themselves whether they were fitted to do the operations, and who tended not to keep detailed records beyond what was required by their own practice.[19]

The AO remains a unique example of a grouping of medical professionals who ran a business to develop surgical products. In some ways they achieved what Charnley could not do – not only were their development processes immaculate and the documentation excellent, but their own product continued to be used only by surgeons who had proved their competence with it. Where that system slotted into academic hierarchies, as in West Germany, the dissemination worked well. In the US it was slow to take off, because the technology seemed foreign and AO was seen as a commercial outfit that was too controlling, and perhaps monopolistic.

By contrast, for THRs in the US, by the 1970s, neither the Charnley/Thackray programme, nor American academic centres, nor professional associations, nor the many competing companies were generally ensuring long-term quality, or data, or the requisite surgical experience. Only a few academic centres behaved differently, notably the Mayo Clinic. Like some of the major UK centres, the Mayo exemplified how professionals could keep excellent long-term records, with good coverage of their localities or regions. But even in Europe, it has usually required government pressure to institute national registries.

Assessment and the roles of government, especially in the UK

We have seen that registers and cost control are important, not just for better expenditure of public monies, but for the better protection of individual patients. British surgeons seem to have been conservative in the use of new devices, in part because the Charnley models served as gold standards. Long-term results for joint replacements outside the major centres would have helped. Infection rates for orthopaedics in a particular hospital have not been assessed as closely as one might have hoped.[20] In this respect, and for obvious reasons, surgery has lagged behind drug treatments, in systematic trials and assessments.

Though schemes were prepared for launching a British register 25 years ago, there was insufficient pressure from industry, government or the professional bodies until after the inquiry into the Capital Hip. This lack of professional commitment is one reason why, in a climate of increased litigation, league tables, risk management and governmental micro-management of services, medical surveillance in the UK has now fallen to various lay managers and assessors rather than to the professions most directly concerned.

We noted that in most countries with registers, they are maintained by professional bodies, even when funded by government. That British professionals do not have ownership of such measures may be the more regrettable when the government is also increasing its controls on the employment of doctors and on professional training and standards.

Professions have developed in areas where public judgement is difficult, whether by patients, public bodies or government. The growth of statistics and randomised controlled trials, and the recent flourishing of information technology, have changed the context of discussion. It is now possible to collect statistics on the performance of devices, hospitals or surgeons, and one should. But it is also the case that such statistics may be misleading, and that one still needs the judgement of experienced professionals who have earned public trust. If such professionals do not feel 'ownership', or a general responsibility for services, then the quality of the data and of the resultant actions is likely to decline.[21]

The best protection is *public-minded* professionals ready to evaluate and reform services as well as particular remedies. We have met many such in this history; but the structures that produced them and supported them, especially the regional and academic medical centres, are under pressure in both the US and the UK.[22] Economists, medical businesses and like-minded politicians now tend to plan medicine as if it were simply a matter of market choices, and professionals were an oddly intractable and expensive form of employee. The dominance of economics in health policy in the UK and the US has marginalised political analysis, or historical perspectives, as it has also marginalised the experience of medical professionals.[23] There is a link here between innovation in health services and in medical technologies. The fashionable confidence in economic principles, which ignores the need for careful testing of organisational forms, curiously echoes the kinds of technical dogmatism, data-shortage and market orientation which sometimes proved unfortunate in THRs.

In the present debates over THR provision in England now, one can see rather clearly these tensions between professional structures, market principles and state power.

Struggling to provide: THRs in England now

In the US, from the 1970s, the medical businesses (from insurance to hospitals, pharmaceuticals to devices) have become at least as influential as the specialist associations and academic leaders. In the UK, and especially since 1997 under New Labour, these same forces (and sometimes the same companies) have grown more powerful, and they are presently favoured by a government understandably keen to see results from the monies which it has poured into the NHS since 2000. This burst of funding has made up much of the deficit which had been apparent from the 1970s when the UK began to be compared with the other rich countries of Europe, but the principles by which the money has been spent since 2000 have drawn on the neoliberal policies of the early 1990s, including American experience. The separation of public purchasing from provision is axiomatic, national targets are set and reinforced by financial penalties, competition between providers is encouraged, and private businesses are encouraged as part of that mix. Much store is now set by new arrangements for patient choice. The special centres for elective surgery, including THRs, are iconic.[24]

The recent improvements of NHS services seem to be chiefly in the provision of new buildings, the shortening of waiting times for emergency accidents and outpatients, the reduction of waiting lists for elective surgery (not least THRs), and the development and co-ordination of services in cancer, heart disease and mental illness. But if you look closely at these improvements, there is little reason to suppose they depended on the repeated structural changes since the early 1990s. Some are the result of much better data, made possible by IT: for example the revelation that the services of the regional centres did not extend across the whole of their regions, and that one needed further mechanisms for spreading best practice.[25] The best of the recent service improvements, for example for cancer, were mostly the result of careful professional planning and increased resources, with the central, professional 'czars'. For these service improvements, the professionals 'bought in'; but that has not been the tone of many of the reforms, including some of the new centres to reduce surgical waiting lists.[26]

As we have seen, private surgical companies became involved in THR provision when the government took the advice of leading surgeons and set up centres that would focus on elective surgery, partly to avoid the disruptions of emergency care, and partly to increase the capacity of the system. This welcome innovation was made controversial when the government pushed private-company centres at the expense of those run from within the NHS. The reasons for this preference appears to be ideological. When compared to the special centres fully within the NHS, the privately run centres were more expensive because of setting-up costs; they did not offer significant improvements in performance, and they did raise problems of co-ordination, quality assurance, and the 'creaming' of easy cases. Their supporters saw the threat of competition as increasing the efficiency of the normal NHS services.[27]

To some extent, the force of complaints from surgeons about the present marketisation and privatisation is limited by public suspicion that their own private practices had benefited substantially from the long-standing waiting lists which the independent special treatment centres (ISTCs) reduced by employing extra surgeons, often imported, working on salaries. But special centres *within* the NHS would seem to have met these goals whilst maintaining better liaison with the rest of the service; and reducing the NHS waiting lists has not yet reduced the private sector demand, as noted in Chapter 6.

These issues are important for the further development of THRs. The problem of separating 'easy surgery' from the rest, and under separate ownership, applies with double force at specialist NHS orthopaedic centres like Wrightington, which fear they will be 'creamed', both by private sector treatment centres and by ordinary NHS hospitals which have lower costs. Regional centres worry that they will be inadequately funded, not just for the difficulty of their remaining cases, but for their research and development functions. In quasi-market models, there is little provision for regional leadership in research and in the development of services, though, as we have seen, these hierarchies were crucial in the NHS through to the 1990s.[28]

That these issues of governance and marketisation are currently intersecting with questions of professional control and quality maintenance is well-illustrated by recent discussion of orthopaedic surgery in Nottingham (a university hospital where Charnley disciples developed THR surgery, including an outstanding database.) The European doctors imported to staff the nearby independent treatment centre are fully qualified, but are accustomed to working under supervision, which they do not now receive; and they are not connected into the regional orthopaedic community. Because the private company which runs the centre has an exclusive contract with a single THR manufacturer, some of the surgeons also find they are working with unfamiliar equipment and prostheses. When the government was asked for a review of the clinical results, they agreed, but then did not provide the data.[29]

Whatever the outcomes here, clinical or indeed political, one could hardly ask for a clearer example of the intersections of technical assessment, professional regulation, service provision and the interests of the political party in power. Paradoxically, perhaps, neo-liberalism and the imitation of business has produced a concentration of power in Downing St which was inconceivable when the NHS began in 1948.

Hospitals, garages and public values

We have followed THRs as illustrating various key aspects of medicine since WWII: from the inventiveness of postwar Britain, to direct-to-consumer advertising in the US; from pioneering NHS surgeons to companies with many of the benefits and costs of the pharmaceutical industry; from the old ladies of Lancashire to the would-(still)-be athletes surfing the web for hip-resurfacings; and from professional control to government regulations and surveillance. Above all, this is a story of how surgical devices which were experimental around WWII became a reliable technique that has hugely benefited millions of people – not by lengthening their lives, but by decreasing pain and increasing mobility in the extra decades of life. Truly the THR is a successful technical fix for one major problem of ageing.

Most aspects of THR reinforce this technical image of medicine more generally. The increasing power of the manufacturers, the specialisation of surgeons, the special treatment centres separated by ownership as well as function, the exclusive deals with companies, and (in the US) the direct-to-patient advertising – all suggest the industrialisation of medicine. Indeed, one could hardly ask for a clearer example of hospital as 'garage', if we may borrow a term from the leading British historian of the recent NHS.

In searching to characterise recent British medicine, Rudolph Klein contrasted the 'garage' with the 'church' model more characteristic of the postwar decades.[30] When Charnley began his work, so the argument would go, hospitals were primarily places of care which expressed community and shared values; now they maintain bodies like garages maintain cars. We have seen that medical services could indeed become more like garages, and not so much the corner garage run by a mechanic as the dealerships run by the major automobile firms.

One model of American hospitals is of a common facility in which individual doctors, and especially surgeons, work as individuals. They work there because the public hospitals offer more investment, more collateral support, and more collective reputation than is easily obtained in the private hospitals which some doctors own. It is not hard to imagine new kinds of hospital facility in which one of the THR companies would run the joint surgery, and one of the dialysis companies the nephrology, etc. – rather like those department stores which house several specialist franchises for jewelry, coats or cosmetics. Indeed, some US health-care analysts, fresh from their previous belief in tariff payments as a major form of cost control, now place their faith in this kind of segmented organisation. Vertical divisions by specialist function may bring economies by providing similar services in many locations (and many countries).[31] Thus the internationalised manufacture of prostheses and other technologies might extend into the provision of particular services. You might become the owner of a DePuy hip, say; you would in some sense be 'branded,' returning to the dealership for your checkups and your revision operation.

One can see the real attractions of this model of care; and one can also see the forces behind it, in the UK as well as the US. The big medical companies are influential with governments, and not just because they offer important international experience and neat remedies for service problems. They are also expected to provide the jobs, exports, and national prestige commonly associated with high-tech business. Indeed, as the older industrial countries try to meet the challenge of newer economies with lower costs, the export of expensive services seems a promising avenue. The World Bank rules about competition are meant to further these developments, not just in poorer countries, but in the richer countries where governments are shifting national health systems towards various forms of market competition. If medical services are becoming more like manufacturing industries, it is partly because governments support the economic well-being of suppliers along with the physical well-being of the people.[32]

The present English moves to 'introduce a variety of suppliers' are presented as a dash of commerce in a public system; but some of these privatisations may be difficult to reverse, and EU competition rules may come to limit the choices open to public representatives. The government has recently moved in circles, but contracts and commercial law are ratchets; and they are bound to prioritise private over public values. Which again brings us back to the need for professional and public values to take precedence.

Grateful as we should be for refined surgical services and complex technologies, we need to remember the crucial elements which separate medicine from trade: that prevention is better than cure, that low-tech treatments may be more effective than high-tech, that public welfare depends on professional restraint, that most of the innovation comes from the public sector, that professional hierarchies are integral to good practice, and that social values also contribute to health. As THRs show rather well, the success of multi-national manufacturers is not to be equated with the public good (nor attributed to their own inventions); nor is high expenditure equivalent to good treatment or wise investment. Even in the US, orthopaedic surgeons have been led to complain of the excessive influence of companies,[33] and their own reputations as professionals are not increased when the companies are brought to court to explain their relations with surgeons.[34]

It is widely recognized that many of the clinical problems which are now solved, or pushed-back, by THRs (and several other expensive procedures) are partly due to life-styles and especially to obesity. Better diets and sensible exercise could greatly reduce the need for THRs, with benefits to private as well as public purses. We can hope that Government agencies, and maybe some companies, will find incentives for direct-to-consumer advertising about prevention, or for promoting increased use of low-tech solutions.

And if technologies such as THR need to be considered in the context of disease prevention and of investment in other potential treatments, so too they need to be provided within health systems alive to wider considerations. We have seen the importance of subjecting new technologies to rigorous, long-term assessment, and of professionals responding to such assessments rather than to market pressures. The AO experiment in Switzerland, like *some* of the recent British experiments in service delivery, may indicate ways of using private companies within frameworks which are determined by those professionals who can take a wider view than companies, or indeed by governments who may be able to draw on various professionals, as well as publics, to take still wider views.

These matters are complicated, and as noted above, the forms of organisation of medicine need testing as much as the drugs or devices. The balances between suppliers, professionals and governments require careful experimentation, rigorously assessed and publicly debated. This is true at the level of particular technologies or for health services more generally; and at various geographical levels, from locality to nation. Part of the the THR story, especially in the US, is of technical and financial exuberance, with relative penury of information. In both the US and the UK, there is now a tendency to assume that change must be driven by commercial or managerial entrepreneurs, and that the consequences can be known in advance. That is true neither for THRs nor for medical services. At both levels we need careful experimentation, devised and assessed by professionals who have tacit and local knowledge, as well as the instruments such as trials and registers which can help bring objectivity.

Regional medical centres used to play a key role in these mediations, but they are being marginalised in England as services are marketised, and national direction replaces regional discussions. This may prove unfortunate, especially if one wants to encourage professionals to help take wider responsibilities for the development of services which are efficient not just in the supply of technical services, but in the maintenance of high standards of health and care. The health of services in given localities benefits from confidence between leading professionals, strategic managers and public representatives with a sense of community. Again, joint treatments can help us see the links between technologies and communities.

It is true that many UK hospitals once functioned like churches, indeed in towns that were divided by class and religion, they were much more effective than churches as agents of social cohesion.[35] Market models of provision, or technocratic models of service distribution, miss this dimension and its continued relevance. Divided communities bring public costs, which maybe substantial, and they do not encourage the mutual assistance which, for example, helps maintain people in their own homes. Whether arthritis necessitates institutional care is in part a matter of surgical remedies, or medicines; but it is also a matter of social support networks which are not easily substituted by professional domiciliary care. How well patients recover depends on the technicalities, but it also depends on truly suppportive doctors and nurses, inside and outside hospitals. And that depends on morale as well as assessments and controls; it depends on a positive and expansive appreciation of professional commitment and collective values.

As British history well shows, hospitals can be strong communities within strong communities, but the community and professional values which carried into the NHS after WWII now seem diminished. The stress on speeding up elective surgery by the importation of surgeons in separate businesses, limits the wider involvement of surgeons at the same time as governments appropriate their professional regulative roles. These moves may accord with the principles of role separation which permeate so much management writing – but role separation is antipathetic to the most publicly useful kinds of professionalism. Surgeons such as Charnley, or Müller in AO, were not just technicians; they were capable of creating both new technologies and new contexts, for the development of procedures and for the improvement of services. And if one looks to the teachers of a Charnley, and to the early years of the NHS, one sees the heavy involvement in local concerns and the responsibilities some doctors took for the improvement of medical services and medical education both across the region and through national politics.[36] That sense of collective endeavour, which can link publics and professionals in constructive dialogue, is a key means of tempering the self-interest of providers, the stoking-up of public demand by corporate interests, and the ignorance of distant government. We have seen that Charnley and his prosthesis were products of a regional medical culture and its wider connections, as AO was a product of Swiss confraternities.

It has been fashionable in health policy, to depict professionals as self interested and to play up the efficiency of market mechanisms, in services as in industry.[37] In many studies of innovation, it is almost axiomatic that large industrial companies provide a range and quantity of financial and technical resources that is imperative for product development. But THRs seem to tell a different story.

THRs were the inventions of surgeons in the early NHS, who worked with engineers and who took responsibility for the procedure as a whole, helping direct the manufacturers. That was how the breakthroughs were produced, and how many innovations are still produced – substantially in the public sector. The diffusion of the THRs worked best where it was led by regional professional centres and nationally assessed. Where these conditions were not present, and especially in the US, business-minded surgeons and large manufacturers could more easily turn hip replacements into high-tech services and corporate markets. The results are of general interest.

For all the innovation as the industry expanded and globalised, many of the resultant THR prostheses (and operations) were worse documented, more expensive and less effective than those produced by the early programmes. And the major later innovation for arthritic hips, the Birmingham hip resurfacing process, came in the 1990s from two British surgeons and their small local company.

Notes

1 Total Hip Replacement: Introduction, Sources and Outline

1 'Railway Engineer's Suicide', *The Times* (9 May 1933).
2 Henderson, Melvin S. and Pollock, George A., 'Surgical Treatment of Osteoarthritis of the Hip Joint', *Journal of Bone and Joint Surgery*, 22 (1940), 923.
3 Cortisone was first synthesised in 1948 by the pharmaceutical company Merck. For the story of cortisone see Le Fanu, James, *The Rise and Fall of Modern Medicine* (London: Little, Brown and Company, 1999), 17–28.
4 Most of the early designs of these single component prostheses were developed mainly in the US and in France. See Coventry, Mark B., 'Lessons Learned in 30 Years of Total Hip Arthroplasty', *Clinical Orthopaedics and Related Research*, 274 (1992), 22–9.
5 See Lerner, Barron H., *The Breast Cancer Wars: Hope, fear and the pursuit of a cure in twentieth-century America* (Oxford: Oxford University Press, 2001). For reflections on the role of surgery in different countries see Pickstone, John V., 'Contested Cumulations: Configurations of cancer treatments through the twentieth century', *Bulletin of the History of Medicine*, 81 (2007), 164–96.
6 For a history of TB in the twentieth century see Bryder, Linda, *Below the Magic Mountain: A social history of tuberculosis in twentieth century Britain* (Oxford: Oxford University Press, 1988).
7 Osteoarthritis is a degenerative condition, usually seen in older people and often affecting weight bearing joints in the lower limbs. See Cantor, David, 'Representing "The Public": Medicine, charity and emotion in twentieth century Britain', in Sturdy, Steve (ed.), *Medicine, Health and the Public Sphere in Britain, 1600–2000* (London: Routledge, 2002), 145–68.
8 *South Bend Tribune* (12 October 2003).
9 *The Financial Times* (22 May 2003).
10 Eftekhar, Nas, 'Founder of the Hip Society (USA) Frank E. Stinchfield', *Clinical Orthopaedics and Related Research*, 285 (1992), 12.
11 Charnley's significant other works include *The Closed Treatment of Common Fractures* (London: Churchill Livingstone, 1950), *Compression Arthrodesis* (London: Churchill Livingstone, 1953).
12 Sarmiento, Augusto, *Bare Bones: A Surgeon's Tale: The price of success in American medicine* (Amherst: Prometheus, 2003).
13 See Faulkner, Alex, 'Casing the Joint: The material development of artificial hips', in Katherine Ott, *et al.* (eds), *Artificial Parts, Practical Lives: Modern histories of prosthetics* (New York: New York University Press: 2002), 199–226.
14 Miller, Dane, 'Orthopaedic Product Technology During the Second Half of the Twentieth Century', in Klenerman, Leslie (ed.), *The Evolution of Orthopaedic Surgery* (London: Royal Society of Medicine Press, 2002), 211–28.

15 Klenerman, Leslie (ed.), *The Evolution of Orthopaedic Surgery* (London: Royal Society of Medicine Press, 2002).
16 Klenerman, Leslie, 'Arthroplasty of the Hip', in Klenerman, Leslie (ed.), *The Evolution of Orthopaedic Surgery* (London: Royal Society of Medicine Press, 2002), 13–23.
17 Heck, Charles V., *1933–1983 Fifty Years of Progress* (Chicago: American Academy of Orthopaedic Surgeons, 1983).
18 Waugh, William, *A History of the British Orthopaedic Association: The first seventy five years* (London: British Orthopaedic Association, 1993).
19 See Jeffrey, Kirk, *Machines in Our Hearts: The cardiac pacemaker, the implantable defibrillator and American health care* (Baltimore: Johns Hopkins University Press, 2001).
20 See Blume, Stuart, *Insight and Industry: On the dynamics of technological change in medicine* (Boston: MIT Press, 1992).
21 See Howell, Joel, *Technology in the Hospital: Transforming patient care in the twentieth century* (Baltimore: John Hopkins University Press, 1995).
22 See Pickstone, John (ed.), *Medical Innovations in Historical Perspective* (Basingstoke: Macmillan, 1992); Lowy, I. (ed.), *Medicine and Change: Historical and sociological studies of medical innovation* (Paris: INSERM/John Libbey, 1993); Stanton, J. (ed.), *Innovations in Health and Medicine: Diffusion and resistance in the twentieth century* (London: Routledge, 2002); Lawrence, G. (ed.), *Technologies of Modern Medicine: Proceedings of a seminar held at the science museum* (London: Science Museum, 1994).
23 See Stanton, J., 'Making Sense of Technologies in Medicine', *Social History of Medicine*, 12 (1999), 437–48; Marks, H.M., 'Medical Technologies: Social contexts and consequences', in Bynum, W. and Porter, R. (eds), *Companion Encyclopaedia of the History of Medicine*, Vol. 2 (London: Routledge, 1993), 1592–618.
24 See Cooter, Roger, 'The Politics of a Spatial Innovation: Fracture clinics in inter-war Britain', in Pickstone, John V. (ed.), *Medical Innovations in Historical Perspective*, 146–64.
25 Sturdy, Steve, 'From the Trenches to the Hospitals at Home: Physiologists, clinicians and oxygen therapy', in Pickstone, John V. (ed.), *Medical Innovations in Historical Perspective*, 104–23.
26 See Cantor, David, 'Cortisone and the Politics of Drama', in Pickstone, John V. (ed.), *Medical Innovations in Historical Perspective*, 165–84.
27 'Hip Histories' at the RCS ran from March to July 2006 and more details can be found on the following websites: http://news.bbc.co.uk/1/hi/health/4906010.stm and http://www.hero.ac.uk/uk/culture_sport/archives/2006/radical_surgery.cfm. 'The Patient's Journey' is a permanent exhibition in the foyer of Wrightington Hospital, opened in October 2006; it was designed to educate patients about the benefits, risks and unknowns of Joint Replacement Surgery. 'The Surgeon's Journey' is an exhibition devoted to the work of John Charnley and the development of hip replacement at Wrightington. It will open in summer 2007 and it will be located in the Biomechanics Laboratory where Charnley carried out his experiments. The Hunterian Museum at the RCS (London) has a small permanent display on THR in its upper gallery on the history of surgery; the Thackray Medical Museum (Leeds) also has a display in its 'Pain, Pus and Blood' gallery, that

explores the relationship between Charnley and the Thackray company. In the US, there are a number of museums of science and technology that cover general themes related to THR, including the Museum of Science in Boston, the Museum of Science and Industry in Chicago, and the museums of the Smithsonian Institution in Washington D.C. In the UK, the Museum of Science and Industry in Manchester and the Science Museum in London both have many exhibitions devoted to technology, including medical technologies. The Science Museum has websites focussing on technology, including http://www.ingenious.org.uk and http://www.makingthemodern-world.org.uk.

28 The transcript of the seminar is published in the volume: Reynolds, L.A. and Tansey, E.M. (eds), 'The Early Development of Total Hip Replacement', in *Wellcome Witnesses to Twentieth Century Medicine*, Vol. 29 (London: The Wellcome Trust Centre for the History of Medicine at UCL, 2007).

29 See respectively Schwartz-Cowan, Ruth, *More Work for Mother: The ironies of household technologies from the open hearth to the microwave* (New York: Basic Books, 1983); Hounshell, David A., *From the American System to Mass Production, 1800–1932* (Baltimore: Johns Hopkins, 1988); and Hughes, Thomas P., *Networks of Power: Electrification in western society 1880–1930* (Baltimore: Johns Hopkins, 1983).

30 Rosenberg, Nathan, *Inside the Black Box: Technology and Economics* (Cambridge, CUP, 1982). The volume edited by Pickstone (1992) originated in a Manchester seminar series intended to bring together history of medicine, history of technology, economics of technology, and innovation and policy studies.

31 See e.g., Metcalfe, J.S., *Evolutionary Economics and Creative Destruction* (London: Routledge 1998).

32 See e.g., Pinch, Trevor J. and Bijker, Wiebe E., 'The Social Construction of Facts and Artefacts: Or how the sociology of science and the sociology of technology might benefit each other', *Social Studies of Science*, 14 (1984): 399–441, and the essays in MacKenzie, Donald and Wajcman, Judy (eds), *The Social Shaping of Technology: How the refrigerator got its hum* (Buckingham: Open University Press, 1985).

33 See the series edited by Geljns, Annetine C., *et al.* entitled *Medical Innovation at the Crossroads*, and published by the National Academy Press, Washington; it discusses medical technologies from an economic rationale.

34 Metcalfe, J.S. and Pickstone, John, 'Replacing Hips and Lenses: Surgery, industry and innovation in post-war Britain', Webster, Andrew (ed.), *New Technologies in Health Care: Challenge, change, innovation* (Basingstoke: Palgrave, 2006), 146–60.

35 Timmermann, C. and Anderson, J. (eds), *Devices and Designs: Medical technologies in historical perspective* (Basingstoke: Palgrave, 2006).

36 Anderson, Julie, 'Greenhouses and Bodysuits: The challenge to knowledge in early hip replacement surgery', in Timmermann, C. and Anderson, J. (eds), *Devices and Designs*, 175–92.

37 Anderson, Julie, 'Innovation and Locality: Hip replacement in Manchester and the Northwest', *Bulletin of the John Rylands Library of the University of Manchester*, Vol. 87 (2007), forthcoming.

38 Cooter, Roger, *Surgery and Society in Peace and War. Orthopaedics and the organization of modern medicine* (Basingstoke: Macmillan, 1993).

39 See Pickstone, John V., *Medicine and Industrial Society: A history of hospital development in Manchester and its region 1752–1946* (Manchester: Manchester University Press, 1985); Cooter, Roger and Pickstone, John, 'From Dispensary to Hospital. Medicine, community and workplace in Ancoats, 1828–1948', *Manchester Regional History Review*, 7 (1993), 73–84; and Butler, S.V.F., 'Academic medicine in Manchester: The careers of Geoffrey Jefferson, Harry Platt and John Stopford', in *Bulletin of the John Rylands Library of the University of Manchester*, Vol. 87 (2007).

40 This long Manchester history, from the bone-setters near Rochdale to the present day connections of Wrightington with multinational prosthesis manufacturers, is explored in Pickstone, John V., 'Bones in Lancashire: Towards long-term contextual analysis of medical technology', in Timmermann, C. and Anderson, J. (eds), *Devices and Designs*, 17–36. It is, throughout, a very transatlantic history.

41 Schlich, Thomas, *Surgery, Science and Industry: A revolution in fracture care, 1950s–1990s* (Basingstoke: Palgrave, 2002).

42 McKee had also visited Thompson in New York and used a modified Thompson early in his practice. Philip Wiles used a two-component hip prosthesis – one of the first known designs. See Wiles, Philip, 'The Surgery of the Osteo-Arthritic Hip', *British Journal of Surgery*, 45 (1958), 488–97.

43 In a few cases it was agreed with Charnley that a prosthesis could be called the Charnley type. These included one by Maurice Müller from Berne. A vast number, however, were not.

44 Stryker fact book, 2006–7 http://library.corporate-ir.net/library/11/118/118965/items/233597/2006_2007FactBook.pdf

45 Hidefjäll, Patrik, 'Biotronik: 40 years of German entrepreneurship in medical technology', in Timmermann, C. and Anderson, J. (eds), *Devices and Designs: Medical technologies in historical perspective* (Basingstoke, Palgrave, 2006) 112–36; Jeffrey, Kirk, *Machines In Our Hearts*; Thomas Schlich, *Surgery, Science and Industry.*

46 See the *National Joint Registry Report Summary* (2004), 6 or http://www.njr-centre.org.uk/documents/reports/annual/1st/sumreport04.pdf

47 Saleh, K.J., *et al.*, 'Economic Evaluations in the Hip Arthroplasty Literature: Lessons to be learned', *Journal of Arthroplasty*, 14:5 (1999), 527–32.

48 Madhok, R., *et al.*, 'Utilization of Upper Limb Joint Replacements During 1972–90: The Mayo Clinic experience', *Proceedings of the Institution for Mechanical Engineers*, Vol. 207 (1993), 239–44.

49 The Royal College of Surgeons of England, *An Investigation of the Performance of the 3M Capital Hip System*, London (2001), 12.

2 Surgeons and Engineers in Postwar Britain: Medical Technology in Local Contexts

1 This procedure is known as 'arthrodesis' (the surgical fusion of a joint to reduce pain and improve stability). It was developed around 1800 by P.F. Moreau in France and H. Park in Liverpool for excising tuberculosis

joints to get rid of the infection (Park, H. and Moreau, P.F., *Cases of the Excision of Carious Joints* (Glasgow: J. Scrymgeour, 1806). It was always a more common procedure for the knee (and later the spine, wrist and ankle) than the hip. The term was first used to describe an ankle operation by Eduard Albert of Vienna in 1882 (see 'Einige Fälle von kunstlicher Ankylosenbildung an paralytischen Gliedmassen', *Wiener medizinische Presse*, 23 (1882), 725–8) and further developed by H.A. Brittain in Norwich, who published *Architectural Principles in Arthrodesis* in 1952.

2 The most common procedure to restore movement in a painful, stiff hip in the early 20th century was 'osteotomy' (the surgical cutting of a bone to shorten, lengthen or realign it). In 1827 the US surgeon, John Rhea Barton, described an operation which cut through the femur between the greater and lesser trochanters to create a false joint to secure motion in an ankylosed hip ('On the treatment of ankylosis by the formation of artificial joints', *North American Medical & Surgical Journal*, 3 (1827), 179–92). In Liverpool in the 1930s, Thomas Porter McMurray developed a displacement osteotomy for fractures of the femoral neck and arthrodesis of the hip (McMurray, T.P., 'Osteoarthritis of the hip joint', *Journal of Bone and Joint Surgery*, 21B (1939), 1–11).

3 Ghormley, Ralph K. and Coventry, Mark B., 'Surgical Treatment of Painful Hips of Adults', *Journal of Bone and Joint Surgery*, 22B (1942), 426.

4 Henderson, Melvin S. and Pollock, George A., 'Surgical Treatment of Osteoarthritis of the Hip Joint', *Journal of Bone and Joint Surgery*, 22B (1940), 930.

5 *The Orthopaedist*, Vol. 1, No. 3 (March 1920), 6.

6 For instance, treatment at the Wrightington TB hospital (near Wigan), opened in 1933, combined established sanatorium regimes and the teachings of Sir Robert Jones (of Liverpool), leading to a scheme of rest, fresh air and good food, augmented with artificial light therapy. No surgery was performed on the infected joint until the disease was cleared. For more details see Swinburn, W.R., *Wrightington Hospital 1933–1983: The story of the first 50 years* (Wigan: Wrightington Hospital, 1983).

7 Wiles reported the results of this pioneering surgery in Wiles, P.W., 'The surgery of the osteoarthritic hip', *British Journal of Surgery*, 45 (1958), 488–97.

8 These half replacements for the femur of the hip are known as 'hemi-arthroplasties' and early ivory prostheses were used by Themistocles Gluck in Berlin (1894) and Hey Groves in Bristol (1923).

9 These procedures are known 'inter-positional arthroplasties'. Louis Ollier in Lyon (1885), John B. Murphy in Chicago (1902) and Putti in Italy (1921) tried a number of inter-positional materials including gold and fat and fascia.

10 Moore, Austin T. and Bohlman, Harold R., 'Metal hip joint. A case report', *Journal of Bone and Joint Surgery*, 25A (1943), 688–92.

11 Judet, J. and Judet, R., 'The Use of an Artificial Femoral Head for Arthroplasty of the Hip Joint', *Journal of Bone and Joint Surgery*, 32B (1950), 166–73.

12 Thompson, F.R., 'Vitallium Intramedullary Hip Prosthesis – preliminary report', *New York State Journal of Medicine*, 52 (1952), 3011–20; Thompson, F.R., 'Two and half years' Experience with the Vitallium Intramedullary Hip Prosthesis', *Journal of Bone and Joint Surgery*, 36A (1954), 489–502.

13 Moore, A.T., 'Metal Hip Joint: A new self-locking Vitallium prosthesis', *Southern Medical Journal*, 45 (1952), 1015–19.

14 Law, W.A., 'Late Results in Vitallium Mould Arthroplasty of the Hip', *Journal of Bone and Joint Surgery*, 44A (1962), 1497.

15 Haboush, E.J., 'A New Operation for Arthroplasty of the Hip based on Biomechanics, Photoelasticity, fast-setting Dental Acrylic, and other considerations', *Bulletin of the Hospital for Joint Diseases*, 14 (1953), 242–77.

16 McKee, G.K. and Watson-Farrar, J., 'Replacement of Arthritic Hips by the McKee-Farrar Prosthesis', *Journal of Bone and Joint Surgery*, 48B (1966), 245.

17 In Charnley's 1953 book, *Compression Arthrodesis*, he advocated a technique called 'central dislocation' of the hip, which was a type of arthrodesis, sometimes using a compression screw. For an evaluation of his new method see Charnley, John, 'Experiences in the Evolution of a New Operation for Osteoarthritis of the Hip Joint', *Journal of Bone and Joint Surgery*, 34B (1952), 506.

18 McKee, G.K., 'Faith healing by a Fellow of the Royal College of Surgeons', *The Mancroft Review*, publisher and date unknown, 1.

19 McKee, G.K., 'Faith Healing', 1.

20 McKee, G.K., 'The Use of Metal in Bone Surgery', *Proceedings of the Royal Society of Medicine*, 50 (1957), 837–40.

21 McKee, G.K., 'Artificial Hip Joint', *Journal of Bone and Joint Surgery*, 33B (1951), 465.

22 McKee, G.K., 'Artificial Hip Joint' (1951), 465.

23 McKee, G.K., 'Developments in total hip joint replacement', *Proceedings of the Institution of Mechanical Engineers*, 181 (1966–7), 85–9.

24 McKee, G.K. and Chen, S.C., 'The Statistics of the McKee-Farrar Method of Total Hip Replacement', *Clinical Orthopaedics and Related Research*, 95 (1973), 26.

25 McKee, G.K. and Chen, S.C. (1973), 29.

26 McKee, G.K. and Chen, S.C. (1973), 30.

27 McKee, G.K., 'Development of Total Prosthetic Replacement to the Hip', *Clinical Orthopaedics and Related Research*, 72 (1970), 101–2.

28 McKee, G.K. and Chen, S.C. (1973), 26–33.

29 K.H. Täger in Germany was a notable example. See Täger, K.H., 'Untersuchungen an Oberflächen und Neogelenkkapseln getragener McKee-Farrar-Endoprothesn', *Archiv für orthopädische und Unfall-Chirurgie*, 86 (1976), 101–13, and Täger, K.H., 'Oberflächen und Neogelenkkapseln getragener McKee-Farrar-Endoprothesen', *Archiv für orthopädische und Unfall-Chirurgie*, 87 (1994), 39–49.

30 Postel, M. and Fayeton, C., 'The McKee-Merle d'Aubigné Prosthesis', in Postel, M., Kerboul, M., Evrard, J. and Courpied, J.P. (eds), *Total Hip Replacement* (Berlin: Springer, 1987), 7–9.

31 Langensköld, A. and Salenius, P., 'Total Replacement of the Hip by the McKee-Farrar Prosthesis. A preliminary report of 81 cases', *Clinical Orthopaedics and Related Research*, 70 (1970), 105.

32 Langenskjiöld, A. and Paavilainen, T., 'Total Replacement of 116 Hips by the McKee-Farrar Prosthesis. A preliminary report', *Clinical Orthopaedics and Related Research*, 95 (1973), 143–50.

33 Chapchal, G. and Müller, W., 'Total Hip Replacement with the McKee Prosthesis. A study of 121 follow-up cases using neutral cement', *Clinical Orthopaedics and Related Research*, 72 (1970), 115–22.

34 Sivash, K.M., 'The Development of a Total Metal Prosthesis for the Hip Joint from a Partial Joint Replacement', *Reconstruction Surgery and Traumatology*, 11 (1969) 53–62.

35 Freeman, P.A., 'McKee-Farrar Total Replacement of the Hip Joint in Rheumatoid Arthritis and Allied Conditions', *Clinical Orthopaedics and Related Research*, 70 (1970), 106–14.

36 Patterson, F.P. and Brown, C.S., 'The McKee-Farrar Total Hip Replacement. Preliminary results and complications of 368 operations performed in five general hospitals', *Journal of Bone and Joint Surgery*, 54A (1972), 257–75.

37 Wilson, P.D., Amstutz, H.C., Czerniecki, A., Salvati, E.A. and Mendes, D.G., 'Total Hip Replacement with Fixation by Acrylic Cement. A preliminary study of 100 consecutive McKee-Farrar prosthetic replacements', *Journal of Bone and Joint Surgery*, 54A (1972), 207–36.

38 Bentley, G. and Duthi, R.B., 'A Comparative Review of the McKee-Farrar and Charnley Total Hip Prostheses', *Clinical Orthopaedics and Related Research*, 95 (1973), 127–42.

39 Dandy, D.J. and Theodorou, B.C., 'The Management of Local Complications of Total Hip Replacement by the McKee-Farrar Technique', *Journal of Bone and Joint Surgery*, 57B (1975), 30–5.

40 Tillberg, B., 'Total Hip Arthroplasty using the McKee & Watson-Farrar Prosthesis. A prospective follow-up study of 327 arthroplasties', *Acta Orthopaedica Scandinavica*, 53 (1982), 104–7.

41 August, A.C., Aldam, C.H. and Pynsent, P.B., 'The McKee-Farrar Hip Arthroplasty. A long-term study', *Journal of Bone and Joint Surgery*, 68B (1986), 520–7.

42 Jantsch, S., Schwagerl, W., Zenz, P., Semlitsch, M. and Fertschak, W., 'Long-term results after Implantation of the McKee-Farrar Total Hip Prostheses', *Archives of Orthopaedic and Trauma Surgery*, 110 (1991), 230–7.

43 Schmalzried, T.P., *et al.*, 'Factors correlating with the Long Term Survival of the McKee-Farrar Total Hip Prosthesis', *Clinical Orthopaedics and Related Research*, 329S (1996), S48–S59.

44 McKee, G.K., *The McKee Method of Total Hip Replacement* (Deloro Stellite Surgical: 1973).

45 Visuri, T. and Koskenvuo, M., 'Cancer Risk after McKee-Farrar Total Hip Replacement', *Orthopaedics*, 14 (1991), 137–42.

46 Jacobsson, S., Djerf, K. and Wahlstrom, O., 'Twenty-year Results of McKee-Farrar versus Charnley Prosthesis', *Clinical Orthopaedics and Related Research*, 329S (1996), S60–S8.

47 Waugh, William, *John Charnley: The man and the hip* (London: Springer-Verlag, 1990).
48 For how Charnley fits into a long tradition of orthopaedics in the North West dating back to the 18th century, see Pickstone, John, 'Bones in Lancashire: Towards long-term contextual analysis of medical technology', in Timmermann, C. and Anderson, J. (eds), *Devices and Designs: Technology and medicine in historical perspective* (Basingstoke: Palgrave, 2006).
49 Charnley, John, *Compression Arthrodiesis* (Edinburgh and London: E&S Livingstone, 1953), which includes a chapter on mechanics by J.A.L. Matheson, and histological observations by S.L. Baker, the local professors respectively of mechanical engineering and rheumatism research.
50 For Charnley's own account of how the Unit was set up see Charnley, John, 'The Development of the Centre for Hip Surgery at Wrightington Hospital', in Swinburn, W.R., *Wrightington Hospital 1933–1983: The story of the first 50 years* (Wigan: Wrightington Hospital, 1983), 36–42.
51 Schlich, Thomas, *Surgery, Society and Industry. A revolution in fracture care, 1950s–1990s* (Basingstoke: Palgrave, 2002).
52 Butler, S.V.F., 'Academic Medicine in Manchester: The careers of Geoffrey Jefferson, Harry Platt and John Stopford', in *Bulletin of the John Rylands Library of the University of Manchester*, Vol. 87 (2007), forthcoming.
53 Charnley, John, 'Total Hip Replacement by Low Friction Arthroplasty', *Clinical Orthopedics and Related Research*, 72 (1970), 7–21.
54 Interview: Mr. Harry Craven, Wrightington, 25/9/2001.
55 See Waugh, *Charnley* (1990), 117, for an account of these early, home-grown manufacturing practices.
56 Waugh, *Charnley* (1990), 117.
57 The rig was a sliding beam with samples of the stainless steel femoral component on it. On the 'heads' were samples of the acetabular material. The rubbing action was loaded to the equivalent of 300 pounds per square inch.
58 The operation was developed by the Oswestry GP, and later Oxford orthopaedic surgeon, Gathorne Robert Girdlestone (Girdlestone, G.R., 'Acute Pyogenic Arthritis of the Hip: An operation giving free access and effective drainage', *The Lancet*, 244 (1943), 419–21). It was developed as an operation to excise infected joints that was often life-saving in the pre-antibiotic era and it followed from the earlier work of many surgeons, including Anthony White who performed joint excision in London in 1822. It involves removing most of the hip joint and capsule and it is still used as a last resort, usually after a number of failed THRs. The joint becomes fixed with fibrous tissue and the patient walks with a stick and two-inch shoe raise. They have a limp but usually no pain.
59 Waugh, *Charnley* (1990), 139–40.
60 Interview: Mr. Harry Craven, Wrightington, 25/9/2001.
61 Waugh, *Charnley* (1990), 122.
62 Interview: Mr. Harry Craven, Wrightington, 25/9/2001.
63 Waugh, *Charnley* (1990), 124.

64 In fact, HDPE turned out to have a higher coefficient of friction than most metal on metal articulations in the lab. It was only under pressure in the body that the coefficient significantly dropped.

65 Charnley, John, 'Total Prosthetic Replacement of the Hip', *Triangle*, 8 (1969), 211–16. For further results, see Charnley, John, 'Total Hip Replacement by Low-Friction Arthroplasty' (1970), and Charnley, John, 'The long term results of Low Friction Arthroplasty of the Hip Performed as a Primary Intervention', *Journal of Bone and Joint Surgery*, 54B (1972), 61–76.

66 Wainwright, Penny, *Opposite the Infirmary: A history of the Thackray company 1902–1990* (Leeds: Medical Museum, 1997), 76–7.

67 For an account of Charnley's trochanteric osteotomy surgical approach and method, see Charnley, John, *Low Friction Arthroplasty of the Hip: Theory and practice* (New York: Springer-Verlag, 1979).

68 Charnley, *Low Friction Arthroplasty (*1979), 347.

69 Scales was a prime mover in setting up this first university department of Biomedical Engineering in Britain. On returning from National Service in the Royal Army Medical Corps, Scales had convinced the Director of the Institute of Orthopaedics, H.J. Seddon, that there was a place for biologically inert plastic materials to replace diseased joints and he led a new department of plastics. In the early 50s, he developed metal on plastic knee prostheses to treat tumour cases and was instrumental in the setting up of a bone tumour unit at the RNOH. As mechanical engineers became interested in his work, he persuaded British Petroleum to second one of its engineers to Stanmore. Scales became the first professor of biomedical engineering in the newly-created department in 1974.

70 Duff-Barclay, I., Scales, J.T. and Wilson, J.N., 'The Development of the Stanmore Total Hip Replacement', *Proceedings of the Royal Society of Medicine*, 59 (1966), 948–51.

71 Oppenheimer, B.S., Oppenheimer, E.T., Stout, A.P. and Eirich, F.R., 'Further studies of Polymers as Carcinogenic Agents in Animals', *Cancer Research*, 15 (1955), 333.

72 Duff-Barclay, Scales and Wilson (1966), 949.

73 Duff-Barclay, Scales and Wilson (1966), 950.

74 Duff-Barclay, Scales and Wilson (1966), 951; Scales, J.T. and Wilson, J.N., 'Some aspects of the development of the Stanmore total hip joint prosthesis', *Reconstruction Surgery and Traumatology*, 11 (1969), 20–39.

75 Wilson, J. N. and Scales, J.T., 'Loosening of Total Hip Replacements with Cement Fixation', *Clinical Orthopaedics and Related Research*, 72 (1970), 145–60.

76 Wilson and Scales (1970), 146.

77 Wilson and Scales (1970), 159.

78 Wilson and Scales (1970), 156.

79 Dobbs, H.S., 'Survivorship of Total Hip Replacements', *Journal of Bone and Joint Surgery*, 62B (1980), 168–73.

80 For more on Stanmore's relations to manufacturers, see Phyllis Hampson's contributions to Reynolds, L.A. and Tansey, E.M. (eds), *The Early Development of Total Hip Replacement*, 'Wellcome Witnesses to Twentieth Century Medicine', Vol. 29 (London: The Wellcome Trust Centre for the History of Medicine at UCL, 2007).

81 Ring, P.A., 'Complete Replacement Arthroplasty of the Hip by the Ring Prosthesis', *Journal of Bone and Joint Surgery*, 50B (1968), 720–31.
82 Interview: K. Drabu, Norwich, November 2001.
83 Ring, P.A., 'Press-Fit Prostheses: Clinical experience', in Reynolds, D. and Freeman, M. (eds), *Osteoarthritis in the Young Adult Hip: Options for surgical management* (Edinburgh: Churchill Livingstone, 1989), 210–32.
84 *Orthopaedic Surgery Catalogue* (Downs Surgical, 1983).
85 Interview: K. Drabu, Norwich, November 2001.
86 Ring, P.A., 'Total Replacement of the Hip Joint. A review of a thousand operations', *Journal of Bone and Joint Surgery*, 56B (1974), 44–58.
87 Ring, P.A., 'Five to Fourteen Year Interim Results of the Uncemented Total Hip Arthroplasty', *Clinical Orthopaedics and Related Research*, 137 (1978), 87–95.
88 Ring, P.A., 'Long Term Results of the Uncemented Hip Replacement', in Noble, J. and Galasko, C.S.B., *Recent Developments in Orthopaedic Surgery* (Manchester: Manchester University Press, 1987), 214–21.
89 Andrew, T.A., Berridge, D., Thomas, A. and Duke, R.N.F., 'Long-term Review of Ring Total Hip Arthroplasty', *Clinical Orthopaedics and Related Research*, 201 (1980), 111–22.
90 Bryant, M.J., Mollan, R.A.B. and Nixon, J.R., 'Survivorship Analysis of the Ring Hip Arthroplasty', *Journal of Arthroplasty*, 6, Supplement (1991), 5–10.
91 Albrecht-Olsen P., Owen-Falkenberg T., Burgaard P. and Andersen, P.B., 'Nine-year follow-up of the Cementless Ring Hip', *Acta Orthopaedica Scandinavia*, 60 (1989), 77–80.
92 Interview: with Dr Clive Lee, Exeter, September, 2002.
93 Interview: with Professor Robin Ling, Exeter, September, 2002.
94 Ling, Robin, 'The Development of the Exeter hip' in Faux, J.C. (ed.), *After Charnley* (Preston: The John Charnley Trust, 2002), 146–69.
95 Ling tried the direct lateral approach during one year but lack of patient mobility and increased postoperative pain caused him to return to the posterior. He also used a wide trochanterotomy from the posterior approach when recovering failed stems.
96 Charnley was never convinced but he did finally visit Exeter in 1982, shortly before his death. He was impressed by the research undertaken by Clive Lee, in conjunction with the Department of Chemistry, on the metabolism of methyl methacrylate. He went in theatre with Ling and was characteristically anxious about the positioning of the femoral component, given what you could see with posterior approach. He found the absence of the greater trochanter until the suturing stage very disorientating. (Interview: Dr Clive Lee, Exeter, September, 2002).
97 Ling, R.S.M., Lee, A.J.C. and Thornett, C.E.E., 'The Collarless Intramedullary Stem', *Journal of Bone and Joint Surgery*, 60B (1978), 137.
98 Interview: with Dr Clive Lee, Exeter, September 2002.
99 Ling, Robin, in Faux, J.C. (ed.) *After Charnley* (2002), 152.
100 'First Annual Report', *National Joint Registry for England and Wales* (September 2004).
101 Waugh, *Charnley* (1990), 93 and 130–1.

3 The Industrial Development of British Hip Replacements and the Charnley/Thackray Relationship

1 Downs Orthopaedic Surgery Catalogue, 1983.
2 The factory on Tabard Street closed in 1981.
3 Down Bros. and Meyer and Phelps Catalogue, 1951.
4 Advertising Section, *Journal of Bone and Joint Surgery*, Vol. 46A (1964), no page number.
5 Down Bros. and Meyer and Phelps Catalogue, 1951.
6 Norman S. Martin had shown his prosthesis at the 1948 meeting of the British Orthopaedic Association in Belfast. Down Bros., Mayer and Phelps Catalogue, 1951.
7 Down Bros., Mayer and Phelps Catalogue, 1961.
8 London Splint Company Catalogue, 1964.
9 Advertising Section, *Journal of Bone and Joint Surgery*, Vol. 54B (1972), no page.
10 Lusterlite Pamphlet, no date c. 1950s.
11 Lusterlite Pamphlet, no date c. 1960.
12 Discussion with Alan Humphries, Thackray Museum, Leeds, June 2006.
13 Downs Catalogue, 1978.
14 Advertising Section, *Journal of Bone and Joint Surgery*, Vol. 54B (1972), no page number.
15 Advertising Section, *Journal of Bone and Joint Surgery*, Vol. 56B (1974), no page number.
16 'By the late 1930s we were coming out of the depression and J.O. Zimmer needed capital to grow on so he brought all of his sales force, which were a total of ten people, to Warsaw and sold thirty percent of the company to them. He forgot to invite the guy from New York. Well there are two stories and J.O. Zimmer had died long before I got involved in orthopaedics so I didn't hear his side of the story, although I've read it in print. But I did hear directly from Frank Siemen who ultimately founded an orthopaedic equipment company. He was the New York distributor and he claims that J.O. Zimmer deliberately invited him a day late. J.O. Zimmer's claim was that typical of a New Yorker he showed up a day late. So I don't know which story, but anyway he didn't get a piece of the action and Mr Siemen who was actually a native New Yorker, and Mr Zimmer never got along very well apparently; and so they agreed that he would go to Europe and establish Zimmer's presence in Europe, which he did, and at the same time registered the Zimmer name in his own name in fifty seven countries and ultimately sold the name back to OEC in the mid-80s.' Interview with Dane A. Miller, September 2002.
17 See History – the JRI Story: www.jri-ltd.co.uk.
18 Howse Catalogue, 1976.
19 *The Times* (26 April 1972).
20 Phyllis Hampson in Reynolds, L.A. and Tansey, E.M. (eds), *The Early Development of Total Hip Replacement*, 'Wellcome Witnesses to Twentieth Century Medicine', Vol. 29 (London: The Wellcome Trust Centre for the History of Medicine at UCL, 2007).

21 For a complete history of the Thackray Company see Penny Wainwright, *Opposite the Infirmary: A history of the Thackray company 1902–1990* (Leeds: Medical Museum, 1997).
22 *Ibid.*
23 Letter from Jose Manuel Del Sol to William Waugh, 6 June, 1988.
24 William Waugh, *John Charnley: The Man and the Hip* (London: Springer-Verlag, 1990), 169–70.
25 'Surgeon and Engineer', *The Lancet* (11 February, 1961), 325–6.
26 Charnley, J., 'Surgeon and Engineer', *The Lancet* (25 February 1961), 449.
27 Letter to John Charnley from W.P. Thackray, 7/5/71. (This and the other Charnley letters referenced below are in the Charnley Papers, Manchester Medical Archive, John Rylands University Library, University of Manchester).
28 Wainwright, *Opposite the Infirmary* (1997), 76.
29 Charnley, 'Surgery of the Hip Joint', *British Medical Journal* (19 March 1960).
30 Advertising Section, *Journal of Bone and Joint Surgery*, 52A (1970), no page.
31 Letter to Charnley from T. Beardmore, 12/12/68.
32 Wainwright, *Opposite the Infirmary* (1997), 79.
33 Thackray Catalogue, 1975.
34 Thackray Catalogue, 1970.
35 Thackray Catalogue, 1970.
36 Advertising Section, *Journal of Bone and Joint Surgery*, 52A (1970), no page.
37 *Charnley Instruments for Low Friction Arthroplasty of the Hip Joint*, Thackray Catalogue, 1969, 3.
38 Thackray Catalogue, 1981.
39 Foreword in Thackray Catalogue, 1975.
40 Thackray Catalogue, 1975.
41 Letter to W.P. Thackray from Charnley, 22/4/68.
42 Letter to Charnley from W.P. Thackray, 27/2/68.
43 Letter from Charnley to J.L.W. Lockhart, 26/1/68.
44 (AAOS on line).
45 Put into 2005 prices using GDP deflator, these indicate increases: taking 1968 as 100, to 162 by 1995, 170 by 1996 and to 355 by 2004. While the rise of 62% from 1968–1995 seems reasonable, the doubling in price between 1995 and 2004 seems high.
46 Advertising Section, *Journal of Bone and Joint Surgery*, 54B (1972), no page.
47 These included Howmedica (1972), Deloro Surgical (1972), Zimmer (1972), OEC (1973) and Downs Surgical Ltd (1975). A Charnley prosthesis modified by Kerboull was advertised by Benoist Girard and Co. in 1975.
48 Letter from T. Beardmore to Charnley, 9/7/68.
49 Letter to P.C. Blackmore, Sales Manager London Splint Company, from John Charnley, 1/7/68.
50 Letter from John Charnley to W.P. Thackray, 6/4/72.
51 Letter to W.P. Thackray from Charnley, 4/12/73.
52 Personal Views on the Failure of Thackray to Develop Their Potential According to my Hopes, 13/1/76. Note in Charnley Archive.
53 Advertising Section, *Journal of Bone and Joint Surgery*, 53A (1971), 108–9.

54 Letter from Charnley to P.C. Blackmore, 1/7/68.
55 Charnley-Müllers were also copied and many Charnley-Müller types were advertised in the *Journal of Bone and Joint Surgery*.
56 Letter to Lockhart from Charnley 24/7/70.
57 This was eventually scrapped when it was realised that Thackray would not be able to gain access to competitors' factories to assess whether or not the patent was being infringed. See correspondence between Tod Thackray and Charnley, May 1971.
58 In 1981, one surgeon purchased some prostheses manufactured by a large rival firm and posted them to Thackray as he felt that the other firm's business practices were unethical. Letter from Charnley to Ron Frank, 21/5/80.
59 Discussion with Christopher Faux, 2007.
60 Interview: Mr Kevin Hardinge, Manchester, 2001.
61 Shelly, P. and Wroblewski, B.M., 'Socket Design and Cement Pressurisation in the Charnley Low-Friction Arthroplasty', *Journal of Bone and Joint Surgery*, 70B (1988), 358–63.
62 Letter to J.D. Boyd from Charnley, 27/10/72.
63 Advertisement, *Journal of Bone and Joint Surgery*, 54B (1972), no page.
64 For example, see Waugh, *John Charnley*, pp. 218–19.
65 In the journals surveyed from the 1960s, there were significantly fewer ads for stainless steel than for titanium or Vitallium.
66 Letter to John Thackray from John Charnley, 5/7/72.
67 Presentation to the Institution of Mechanical Engineers, 3/1/74.
68 Letter to John Thackray from Charnley, 9/6/71.
69 Discussion with Alan Humphries, Thackray Museum, Leeds, June 2006.
70 Letter to R. Frank from Charnley, 28/8/75.
71 Letter from Charnley to John Thackray, 27/8/71.
72 Charnley was not actually sued throughout his medical career. Letter to G. Robinson from Charnley, 14/5/73.
73 Report to G.J. Robinson from G. Parkin, 5/11/73.
74 Letter to W.P. Thackray from Charnley, 5/3/74.
75 Letter to W.P. Thackray from Charnley, 14/3/74.
76 Letter to Charnley from M.P. Jones, 3/1/74.
77 Letter to Charnley from J.D. Boyd, 9/4/74.
78 Letter to J.P. Thackray from Charnley, 5/1/71.
79 Letter to Charnley from R. Frank, 23/9/74.
80 Letter to W.P. Thackray from Charnley, 23/11/72.
81 Waugh, *Charnley*, p. 216.
82 OEC was bought by Biomet in the mid-1980s.
83 Company History Listings, Thackray Museum Archive, Leeds, England.
84 www.corin.co.uk/aboutus.asp.
85 *The Sheffield Star* (9 July 2003).
86 www.smithandnephew.com (accessed 15 August 2004).
87 www.smithandnephew.com (accessed 15 August 2004).

4 Medical Technology and Industrial Dynamics: The United States and the Global Market

1 Medical Industry E-Mail News Service (27 June 2005).
2 The Stryker frame was a bed which was designed in the 1930s in which the patient could be turned so as to prevent pressure sores that could become seriously infected. See Breneman, James C., *The Stryker Story: Homer's Iliad* (Michigan: Phil Schubert and Associates, 1992).
3 www.depuy.com (30 May 2003).
4 Zimmer, *In the Spirit of Excellence* (Warsaw: Zimmer, 1986), 7.
5 See Guttmann, Ludwig, *Spinal Cord Injuries: Comprehensive management and research* (Oxford: Blackwell Scientific Publications, 1973), 5; and Ward, R.O., 'The Management of the Bladder in Spinal Injuries', in H. Hamilton Bailey (ed.), *Surgery of Modern Warfare* (Edinburgh: E & S Livingstone, 1944), 661.
6 Advertisement, *Journal of Bone and Joint Surgery*, Vol. 54A (1972), no page.
7 Austin Moore and Harold Bohlman inserted the first Vitallium endoprosthesis in 1943. F.R. Thompson of New York inserted his first prosthesis in 1953. Both prostheses are still in use today. See Klenerman, Leslie, 'Arthroplasty of the Hip', in Klenerman, Leslie (ed.), *The Evolution of Orthopaedic Surgery* (London: The Royal Society of Medicine Press, 2002), 14.
8 Catalogue from the London Splint Company, 1964, no page. The London Splint Company was closely allied with Howmedica.
9 Zimmer, *In the Spirit of Excellence*, 12.
10 Zimmer pamphlet, no date.
11 Advertisements, *Journal of Bone and Joint Surgery* (1966A, 1967A and 1967B), no page numbers.
12 Advertisement, *Journal of Bone and Joint Surgery*, 46A (1964), no page.
13 *Journal of Bone and Joint Surgery*, Vol. 46A, 1964.
14 Advertisement, *Journal of Bone and Joint Surgery*, 46A (1964), no page.
15 Advertisement, *Journal of Bone and Joint Surgery*, 51A (1970), no page.
16 Advertisement, *Journal of Bone and Joint Surgery*, 53A (1971), no page.
17 Advertising section, *Journal of Bone and Joint Surgery*, 56A (1974), no page.
18 Advertisement, *Journal of Bone and Joint Surgery*, 54B, no page.
19 Interview: William Harris, Boston, 2003.
20 Keller, Larry, 'High Court Upholds $20M in Punitive Damages to Palm Beach Gardens Surgeon-Inventor', *Broward Daily Business Review* (12 December 2001).
21 See Breneman, James C., *The Stryker Story* (1992).
22 www.bioproimplants.com (accessed 23 July 2005).
23 Advertisement, *Journal of Bone and Joint Surgery*, 62B (1980), no page.
24 Advertisement, *Journal of Bone and Joint Surgery*, 54B (1972), no page.
25 Zimmer, *In the Spirit of Excellence*, 13.
26 www.biomet.com (accessed October 2005).
27 *The Financial Times* (7 August 2003).
28 In the 1990s, MMT was the market leader in hip resurfacing. Started by two orthopaedic surgeons from Birmingham, Derek McMinn and Ronan Treacy, it was purchased by Smith and Nephew in 2004. See: http://www.smith-nephew.com/who/history-2000s.html.

29 Standard and Poors, 2003.
30 Standard and Poors, 2003.
31 See Mansfield, Edwin, 'Industrial Innovation in Japan and the United States', *Science* (30 September, 1988), 1769–1974. The article compares the R&D budgets of American and Japanese companies.
32 *The Neikki Weekly* (5 November 2002).
33 Interview: Pablo Gomez, Manchester, 2005.
34 Biomed News, Press Release: *2003 Medical Industry Growth to be Fuelled by Major Medical Product Launches* (15 February 2003), biomednews@aol.com.
35 *The New York Times* (18 September 1994).
36 Feder, Barnaby J., 'In the Land of the Orthopaedic Implant', *New York Times* (18 September 1994).
37 Sarmiento, Augusto, *Bares Bones: A Surgeon's Tale* (Amherst: Prometheus Books, 2003), 279.
38 Interview: Jorge Galante, Chicago, November 2002.
39 Interview: William Harris, Boston, 2003.
40 Bentkover, Judith D., *et al.*, 'Case Study #14: Cost benefit/Cost effectiveness of Medical Technologies: A Case Study of Orthopaedic Joint Implants', *The Implications of Cost Effectiveness Analysis of Medical Technology*, Office of Technology Assessment, Washington (September 1981), 7.
41 Interview with Jorge Galante, Chicago, November 2002.
42 One engineer said that more surgeons were involved in each design than before. Interview with Roy Crowninshield, Warsaw, 2002.
43 Interview with Allan Gross, Chicago, 2002.
44 Interview with Wayne Paprosky, Chicago, 2002.
45 Hubbard, Richard F. and Rodengen, Jeffrey L., *Biomet: From Warsaw to the world* (Fort Lauderdale: Write Stuff Enterprises, 2002), 35–6.
46 Interview with Wayne Paprosky, Chicago, 2002.
47 Advertising Section *Journal of Bone and Joint Surgery*, 77A (1995), no page.
48 Interview with Dane A Miller, Warsaw, 2002.
49 Dhillon, B.S., *Medical Device Reliability and Associated Areas* (New York: CRC Press, 2000), 3.
50 Interview with Dane A. Miller, Warsaw, 2002.
51 The advertisement also featured a photograph of the four 1969 winners. *Journal of Bone and Joint Surgery*, 51A (1969), 6.
52 Joint Advertisement OREF and Zimmer, *Journal of Arthroplasty*, 17, Supplement 1 (2002), no page number.
53 These courses are sponsored by Zimmer but conducted by orthopaedic surgeons. Discussion with Brad Bishop, Warsaw, 2002.
54 Anderson in discussion with Orthopaedic Surgeons, AAOS Meeting, New Orleans, 2003.
55 Murray, D.W., *et al.*, 'Which Primary Hip Replacement?', *Journal of Bone and Joint Surgery*, 77 (1995), 527.
56 Homsy, C.A., 'R&D and manufacturing of biomaterials and implants in the socio-political context', in Kossowsky, R. and Kossovsky, N. (eds), *Advances in Materials Science and Implant Orthopaedic Surgery* (Dordrecht: Kluwer Academic Publishers, 1995).
57 *The Guardian* (5 May 2004).
58 *The Guardian* (5 May 2004).

59 *The Guardian* (25 May 2004).
60 Hearings Before the Subcommittee on Public Health and Environment of the Committee on Interstate and Foreign Commerce House of Representatives on Bills H.R. 6073 and H.R. 9984, Amendments to the Food, Drug and Cosmetic Act (23–24 October 1973).
61 Dane A. Miller, http://www.orthosupersite.com.
62 See Greene, Jeremy A., 'Attention to "Details": Etiquette and the Pharmaceutical Salesman in Postwar America', *Social Studies of Science*, 34 (2004), 271–92.
63 J.V. Pickstone, experience in a Canadian physiology lab in mid-60s.
64 Information from Christopher Faux for UK.
65 *The New York Times* (2 November 1977).
66 *The New York Times* (30 October 1977).
67 *The New York Times* (18 December 1977).
68 *The New York Times* (21 May 1978).
69 In observing hip replacement surgery in the United States, Anderson witnessed the company representative with cases full of instruments and implants, should the surgeon require them.
70 Interview by Anderson of AM01, AM02, AE01 and AM03, Warsaw, 2002.
71 AM01, Warsaw, 2002.
72 Interview: Allan Gross, Chicago, 2002.
73 Hubbard and Rodengen, *Biomet* (2002), 80.
74 William L. Healy, Economic Considerations in Total Hip Arthroplasty and Implant Standardisation', *Clinical Orthopaedics and Related Research*, 311 (1995), 102.
75 Healy, William L., 'Economic Considerations' (1995), 102. And the interview with Biomet employees (n. 70).
76 Interview: Wayne Paprosky, Chicago, 2002.
77 Angell, Marcia, *The Truth About the Drug Companies: How they deceive us and what to do about it* (New York: Random House, 2004).

5 Change and Stability in Technical Systems: Materials and Environments

1 Although very recently this has become less of a factor as surgeons at specialist centres are often trained in multiple approaches and select prostheses and approaches based on the clinical needs of the patient.
2 Bijker and his colleagues have looked at competing technologies in the bicycle and concluded that many of the reasons for a particular technology 'winning out' over another were that certain bicycles came to embody the interests of particular social groups. See Bijker, W.E., Hughes, T.P. and Pinch, T. (eds), *The Social Construction of Technological Systems: New directions in the sociology and history of technology* (Cambridge, MA: MIT Press, 1989); Bijker, W.E., *Of Bicycles, Bakelites, and Bulbs: Toward a theory of sociotechnical change* (Cambridge, MA and London, England: MIT Press, 1995).
3 In discussion with Christopher Faux, he suggested that the general consensus on follow up is 6.5 years.

4 An example of a relatively slow feedback systems on failures and the many variables involved can be found in the case of the Charnley Elite-Plus hip system put on the market by DePuy in 1993. Some combinations offered a new type of polyethylene (Hylamer) and a new type of ceramic head (Zirconia). Despite performing well in laboratory tests, 17% of these combinations failed and a further 41% were loose after ten years. But early failures did not become apparent until 1997 and the extent of the failures and potential future failures were not fully know until 2005 after long-term radiostereometric analysis. See Livingston, B.J., *et al.*, 'Complications of Total Hip Arthroplasty Associated with the use of an Acetabular Component with a Hylamer Liner', *Journal of Bone and Joint Surgery*, Vol. 79A (1997), 1529–38; Hauptfleisch, J., *et al.*, 'The Premature Failure of the Charnley Elite-Plus Stem', *Journal of Bone and Joint Surgery*, Vol. 88B (2006), 179–83.

5 For recent examples, see Skurla, C.P., *et al.*, 'Assessing the Dog as a Model for Human Total Hip Replacement: Analysis of 38 canine cemented femoral components retrieved at post-mortem', *Journal of Bone and Joint Surgery*, 87B (2005), 120–7; and Tepic, S., Remiger, A.R., *et al.*, 'Strength Recovery in Fractured Sheep Tibia Treated with a Plate or an Internal Fixator: An Experimental Study with a Two-Year Follow-up', *Journal of Orthopedic Trauma*, 11 (1997), 14–23.

6 See McKee, G.K., 'Development of Total Prosthetic Replacement of the Hip', *Clinical Orthopaedics and Related Research*, No. 72 (1970), 85–103.

7 Scales, John T. and Wilson, J.N., 'Some Aspects of the Development of the Stanmore Total Joint Prosthesis', *Reconstructive Surgery and Traumatology*, Vol. 11 (1969), 20–39.

8 Advertising pages, Zimmer Orthopaedic Limited, *Journal of Bone and Joint Surgery*, 46B (1964), no page.

9 Müller, Maurice M., 'The Benefits of Metal-on-Metal Total Hip Replacements', *Clinical Orthopaedics and Related Research*, No. 311 (1995), 55.

10 Amstutz, Harlan C. and Grigoris, Peter, 'Metal on Metal Bearings in Hip Arthroplasty', *Clinical Orthopaedics and Related Research*, No. 329S (1996), S28.

11 Müller, Maurice M., 'The Benefits of Metal-on-Metal Total Hip Replacements', *Clinical Orthopaedics and Related Research*, No. 311 (1995), 56.

12 Robert Scott, 'Metal on Metal, in Sedel, L. and Cabanela, M.E. (eds), *Hip Surgery: Materials and Developments* (London: Martin Dunitz, 1998), 27.

13 Dorr, *et al.*, 'Modern Metal on Metal Articulation for Total Hip Replacements', *Clinical Orthopaedics and Related Research*, No. 333 (December 1996), 109.

14 Müller, Maurice M., 'The Benefits of Metal-on-Metal Total Hip Replacements', *Clinical Orthopaedics and Related Research*, No. 311 (1995), 58.

15 See Black, J., 'Metal on Metal Bearings: A practical alternative to metal on polyethylene total joints', *Clinical Orthopaedics and Related Research*, Supplement, No. 329 (1996), 244–55.

16 Dorr, *et al.*, 'Modern Metal on Metal Articulation for Total Hip Replacements', *Clinical Orthopaedics and Related Research*, No. 333 (1996), 115.

17 Lerouge, Sophie, *et al.*, 'Alumina Ceramic in Total Joint Replacement', in Sedel, L. and Cabanela, M.E. (eds), *Hip Surgery: Materials and developments* (London: Martin Dunitz, 1998), 31.
18 Pascal, Christel, 'Ceramics for Joint Replacement', in Morrey, B.F. (ed.), *Biological, Material and Mechanical Considerations of Joint Replacement* (New York: 1993), 303.
19 Sandhaus, S., *British Patent Number 1083769*, 1965.
20 Boutin, P., 'Arthroplastie Totale de la Haunche Par Prosthèse en Alumine Fritée, Etude expérimentale et premières applications cliniques', *Review Chir. Orthop.*, No. 58 (1972), 229–46.
21 Advertising Section, *Journal of Bone and Joint Surgery*, 77A (1995), no page number.
22 Lerouge, Sophie, *et al.*, 'Alumina Ceramic in Total Joint Replacement', in Sedel, L. and Cabanela, M.E. (eds), *Hip Surgery: Materials and developments* (London: Martin Dunitz, 1998), 31.
23 Yamamuro, T., 'Zirconia Ceramic for the Femoral Head of a Hip Prosthesis', in Sedel, L. and Cabanela, M.E. (eds), *Hip Surgery* (1998), 41.
24 Willmann, G. and Zweymuller, K., *Bioceramics in Hip Joint Replacement* (2000), 156.
25 Fackelman, Kathleen, 'Hip Implants Get the Active Back in Gear', *USA Today* (24 June 2003).
26 'Stryker Howmedica Osteonics Receives Approval of Ceramic Insert for Hip Replacements', *PR Newswire* (3 February 2003).
27 FDA, 'Medical Devices Recalls of Zirconia Ceramic Femoral Heads for Implants', http://www.fda.gov/cdrh/recalls/zirconiahip.html.
28 Miller, Dane A., 'Orthopaedic Product Technology During the Second Half of the Twentieth Century', in Klenerman, L. (ed.), *The Evolution of Orthopaedic Surgery* (London: RSM Press, 2002), 212.
29 Down Bros. Catalogue (1951).
30 Lusterlite Catalogue (1951).
31 While cemented cups could be difficult to remove from the acetabulum, failed press-fit cups could also make for difficulties. They were so well fixed that their removal could damage the bone; impaction bone grafting was often then required to reconstruct the acetabulum before a new cup could be inserted. Discussion with Christopher Faux, Manchester, 2007.
32 Miller, 'Orthopaedic Product Technology' (2002), 222.
33 Ritter, Merrill A., 'The Cemented Acetabular Component of a Total Hip Replacement', *Clinical Orthopaedics and Related Research*, No. 311 (1995), 69.
34 Postak, P., *et al.*, *Highly Cross-linked Polyethylene Modular Acetabular Designs: Performance characteristics*, Orthopaedic Research Laboratories, Ohio (2003), 4.
35 Wright, Timothy M., 'Discussion of Ultrahigh Molecular Weight Polyethylene', in Morrey, B.F. (ed.), *Biological, Material and Mechanical Considerations of Joint Replacement*, New York (1993), 127.
36 Credit for this cross-linking technology was given to a research team at J. Vernon Luck MD Orthopaedic Research Centre in Los Angeles The product was licensed to DePuy and called Marathon Cross Linked Polyethylene. *Business Wire* (27 March 2002).

37 Ege, W. and Kuhn, F.D., *Industrial Development of Bone Cement* (2001), 7.

38 See Haboush, E., 'A New Operation for Arthroplasty of the Hip Based on Biomechanics, Photoelasticity, Fast Setting Dental Acrylic and Other Considerations', *Bulletin of Hospital Joint Disease*, 14 (1953), 242–77; Wiltse, L.L. and Hall, R.H., 'Experimental Studies Regarding the Possible Use of Self-Curing Acrylic in Orthopaedic Surgery', *Journal of Bone and Joint Surgery*, 39A (1957), 961–72.

39 Bentkover, Judith D., & Drew, Phillip G., *Cost Benefit/Cost Effectiveness of Medical Technologies: A case study of orthopaedic implants*, OTA Background Paper, Cambridge (1981).

40 *The New York Times* (30 November 1971).

41 Lettin, Alan, *Was It Something I Said?* (London: Memoir Club, 2005).

42 Letter to William Waugh from Professor J. Stevens, 22/2/89.

43 Villar, R.N., 'Non-cemented Hip Prostheses', in Letters, *British Medical Journal*, Vol. 292 (6 July 1985), 53.

44 Kuhn, K.D. and Ege, W., 'Composition and Properties', Paper presented at the *Acrylic Bone Cement in the New Millennium Sesquicentennial Conference* (13–14 May 2002).

45 See Lidgren, *et al.*, 'Strength of Polymethylmethacrylate Increased by Vacuum Mixing', *Acta Orthop Scand*, Vol. 55, No. 5 (1984), 536–41; Postak, Paul D., *et al.*, *Assuring Cement fixation: All mixing systems are not the same*, Orthopaedic Research Laboratories, Ohio (2003); Department of Health, *The Control of Substances Hazardous to Health: Guidance for the initial assessment in hospitals* (London: HMSO, 1989), 109. For more on the merits of vacuum mixing see, http://bonecement.com/index.php?id=16942&no_cache=1

46 'New Technique Found for Hip Replacements', *New York Times* (October 24, 1984); Burke, *et al.*, 'Centrifugation as a Method of Improving Tensile and Fatigue Properties of Acrylic Bone Cement', *Journal of Bone and Joint Surgery*, 66A (1984), 1265.

47 http://news.bbc.co.uk/1/hi/health/3180436.stm.

48 Interview William Harris, *Orthopaedics Today* (May 1998).

49 M. Ritter, 'The Cemented Acetabular Component of a Total Hip Replacement', *Clinical Orthopaedics and Related Research*, No. 311, February 1995, p. 73.

50 *The New York Times* (19 September 1994).

51 See Powell, J.N., *et al.*, 'Cardiac Arrest Associated with Bone Cement', *British Medical Journal* (8 August 1970), 326. For discussion on the subject see also Letters, *British Medical Journal* (22 August 1970), 465; (29 August 1970), 523; (5 September 1970), 588; (19 September 1970), 710; (6 May 1972), 350; (27 May 1972), 528; (17 June 1972), 713.

52 Herndon, James H., *et al.*, 'Fat Embolism During Total Hip Replacement: A prospective study', *The Journal of Bone and Joint Surgery*, 56A (1974), 1350.

53 See Jones, L.C. and Hungerford, D.S., 'Cement Disease', *Clinical Orthopaedics and Related Research*, No. 225 (1987), 192–206.

54 See Walczak, *et al.*, 'Low Friction Arthroplasty with Boneloc Bone-Cement Outcome at 2 to 4 Years', *The Journal of Arthroplasty*, 15 (2000) 205–9; Abdel-Kader, 'Boneloc Bone-Cement Experience in Hip

Arthroplasty during a 3 Year Period', *The Journal of Arthroplasty*, Vol. 16, No. 7, 811–19; Espehaug, *et al.*, 'The Type of Cement and Failure of Total Hip Replacements', *Journal of Bone and Joint Surgery*, Vol. 84B (2002), 832–8; Harper, Elizabeth, 'Mechanical Properties', Paper presented at the *Acrylic Bone Cement in the New Millennium Sesquicentennial Conference* (13–14 May 2002).

55 Huo, Michael H. and Salvati, Eduardo A., 'Hip Wear: Clinical Aspects', in Morrey, *Biological Material and Mechanical Considerations of Joint Replacement* (New York: 1993), 245.

56 From 1998–2005 THR is distinguished by whether cemented or not. The totals, shown in Table 1, indicate that although cemented remained the most common type in 2005, a strong trend to non-cemented is apparent. Only 12.5 were non-cemented in 1998 but 26% by 2005.

57 Spector, Myron, 'Historical Review of Porous Coated Implants', *The Journal of Arthroplasty*, Vol. 2, No. 2 (June 1987), 164.

58 Lord, G. and Bancel, P., 'The Medreporic Cementless Total Hip Arthroplasty: New experimentation data and a seven year clinical follow-up study', *Clinical Orthopaedics and Related Research*, No. 176 (1983).

59 Hamblen, D.L. and Paul, J.P., 'The Integrity of Porous Coatings for Cementless Implants', *The Journal of Bone and Joint Surgery*, 70B (1988), 521.

60 Romagnoli, Sergio, 'Press-Fit Hip Arthroplasty: A European alternative', *The Journal of Arthroplasty*, 17, Suppl. 1 (2002), 109.

61 Sivash, K.M., *Alloplasty of the Hip Joint* (Moscow: Central Institute for Traumatology and Orthopaedics, 1967).

62 Mittelmeier, H., 'Ceramic Prosthetic Devices', in Welch, R.B. (ed.) *Proceedings of the Hip Society* (St Louis: 1984), 321.

63 Struthers, A.M., 'An Experimental Study of Polyvinyl Sponge as a Substitute for Bone', *Plastic Reconstructive Surgery*, 15 (1955), 274–89.

64 Mandarino, M.P. and Salvatore, J.E., 'Polyurethane Polymer (Ostamer): Its Uses in Fractured and Diseased Bones', *Surgical Forum*, 9 (1959), 762–5.

65 Smith, L., 'Ceramic-Plastic Material as a Bone Substitute', *Archives of Surgery*, No. 87 (1963), 653–61; Spector, Myron, 'Historical Review of Porous Coated Implants', *The Journal of Arthroplasty*, 2 (1987), 165.

66 Petersen, C.D., *et al.*, 'Union between Bone and Implants of Open Pore Ceramic and Stainless Steel: A histological study', *Journal Bone and Joint Surgery*, 51A (1969).

67 Interview: Jorge Galante, Chicago, 2003.

68 See Closhisy, John C. and Harris, William H., 'The Harris-Galante Uncemented Femoral Component in Primary Total Hip Replacement at 10 Years', *The Journal of Arthroplasty*, 14 (1999), 915–17; Lachiewicz, Paul F., *et al.*, 'A Prospective Study of 100 Consecutive Harris-Galante Porous Total Hip Arthroplasties', *The Journal of Arthroplasty*, 7 (1992), 519–26.

69 Callaghan, John J. (Major), *et al.*, The Uncemented Porous Coated Anatomic Total Hip Prosthesis, *Journal of Bone and Joint Surgery*, 70A (1988), 337–46.

70 The Queen Mother was fitted with Furlongs although they were the standard cemented variety, in 1995 and 1998 respectively.

71 Amstutz, Harlan, *et al.*, 'The Cemented T-28/TR-28 Prosthesis', in Finerman, *et al.* (eds), *Total Hip Arthroplasty Outcomes* (1998), 55.

72 Advertising Section, *Journal of Bone and Joint Surgery*, 75A (1993), no page number.
73 Gross, Allan E. and Hutchison, Carol R., 'Proximal Femoral Allografts for Reconstruction of Bone Stock in Revision Arthroplasty of the Hip', *The Orthopaedic Clinics of North America*, Vol. 29, No. 2 (April 1998), 317.
74 Hamilton, William G., *et al.*, 'Massive Acetabular Impaction Grafting: Mid Term Results', Pamphlet AAOS Congress, New Orleans (2003).
75 AAOS, Committee on Patient Safety, *Musculoskeletal Allograft Tissue Safety Pamphlet* (2003).
76 Video: *Bone Allografts for Surgical Implantation* (BOA, 1994).
77 AAOS and AATB, Pamphlet, *What Can You Tell Me About Bone and Tissue Transplantation?* Rosemont and McLean (2003), 3.
78 'Proceedings and Reports of Councils and Associations: British Orthopaedic Association', *Journal of Bone and Joint Surgery*, 55B, No. 4 (November 1973), 880.
79 Committee on Patient Safety, Committee on Biological Implants, Tissue Work Group, *Musculoskeletal Allograft Tissue Safety Information Pamphlet* (2003).
80 Committee on Patient Safety, Committee on Biological Implants, Tissue Work Group, *Musculoskeletal Allograft Tissue Safety Information Pamphlet* (2003).
81 Stanisavljevic, Stanko, *et al.*, 'Autologous Blood Transfusion in Total Joint Arthroplasty', *The Journal of Arthroplasty*, Vol. 1, No. 3 (September 1986), 208.
82 Waugh, William, *John Charnley: The man and the hip* (London, 1990), 106.
83 See McMinn, D., *et al.*, 'Metal on Metal Surface Replacement of the Hip', *Clinical Orthopaedics and Related Research*, Vol. 329S (1996), 89–98; Treacy, R., *et al.*, 'Birmingham Hip Resurfacing Arthroplasty: A minimum follow up of five years', *Journal of Bone and Joint Surgery*, Vol. 87B (2005), 167–70.
84 For an article on the sale of the company MMT: see http://www.real-business.co.uk/ARTICLE/UNSPUN-Doing-deals/1103.aspx.
85 See Berger, Richard A., 'Total Hip Arthroplasty using the Minimally Invasive Two-Incisions Approach', *Clinical Orthopaedics and Related Research*, Vol. 417 (2003), 232–41; Schofield, S.F., 'Minimally Invasive Surgery for Hip Replacement', *ANZ Journal of Surgery*, 74, No. 4 (2004), 287–8.
86 See Lawler, M., *et al.*, 'Comparison of Early Postoperative Functional Levels Following Total Hip Replacement using Minimally Invasive versus Standard Incisions: A prospective randomised blinded trial', *Clinical Rehabilitation*, Vol. 19 (2005), 465–74; Ogonda, L., *et al.*, 'A Minimal-Incision Technique in Total Hip Arthroplasty Does Not Improve Early Postoperative Outcomes', *Journal of Bone and Joint Surgery*, Vol. 87A, No. 4 (2005), 701–10.
87 http://www.uchospitals.edu/news/1998/19981017-asa-wakeup.html.
88 Results from a study conducted by the Mayo Clinic. *The Lancet* (4 March 1978), 482.
89 Evans, E.M., 'Metal Sensitivity as a Cause of Bone Necrosis and Loosening of the Prosthesis', *Journal of Bone and Joint Surgery*, 56B (1974), 626. *The Lancet* (September 8, 1984), 557.

90 *British Medical Journal* (29 July, 1977), 213.

91 Charnley, J., Letter to the Editor, *The Lancet* (13 July 1968), 101.

92 Scott, C.C., 'Laminar/Linear Flow System of Ventilation', *Lancet*, 295 (1970), 989–93.

93 See Anderson, J., 'Innovation and Locality: Hip replacement in Manchester and the north west of England', *Bulletin of the John Rylands University Library*, 87 (2007), forthcoming.

94 Selwyn, S., 'Airing Operating Theatres', *British Medical Journal*, 327 (1986), 1544–5.

95 In later years the greenhouse had four people inside. Discussion with Christopher Faux, Manchester, 2007.

96 *Proposals for a Sterile Operating Enclosure at the Centre for Hip Surgery, Wrightington*, UK National Archives, 21/1/64, FD 23/616.

97 *Proposals for a Sterile Operating Enclosure at the Centre for Hip Surgery, Wrightington*, UK National Archives, 21/1/64, FD 23/616.

98 Letter from John Charnley to the Medical Research Council, UK National Archives, 14/3/73, FD 23/616.

99 Charnley, J. and Eftekhar, N., 'Post-operative Infection in Total Prosthetic Arthroplasty of the Hip', *British Journal of Surgery*, 56(9) (1969), 641–9.

100 Advertising Leaflet Howorth Systems, Medical Collection, John Rylands Library, University of Manchester (undated: presumed 1976).

101 International Installation list Charnley-Howorth Air Systems, Medical Collection, John Rylands Library, University of Manchester (undated presumed 1989).

102 Advertisement Section, *Journal of Bone and Joint Surgery*, 53A(7) (October 1971), no page number.

103 Scott, 'Laminar/Linear Flow System of Ventilation', 989.

104 Duguid, J.P. and Wallace, A.T., 'Air Infection from Dust Liberated from Clothing', *Lancet*, 252 (1948), 845.

105 Charnley, J. and Eftekhar, N., 'Penetration of Gown Material by Organisms from the Surgeon's Body', *Lancet*, 293 (1969), 173.

106 Charnley found that altering the style of the gown to the pull-on or cape style and using the suction with the new style of gown made a significant difference in the bacterial content of the air. Charnley, J., 'Operating-Room Conditions', Letter to the Editor, *The Lancet*, 286 (1965), 907–8.

107 Whyte, W., *et al.*, 'The Reduction of Bacteria in the Operating Room through the Use of Non-Woven Clothing', *British Journal of Surgery*, 65(7) (1978), 469.

108 See Paine, C.J., 'The Aetiology of Puerperal Infection with special reference to Droplet Infection', *British Medical Journal*, 1 (1935), 243–6.

109 Charnley, J., 'Clean Air Operating Room Enclosures', *British Medical Journal*, 1 (1974), 224.

110 Charnley, J., 'Wound Closure', *Centre for Hip Surgery, Wrightington Hospital Internal Publication No. 33* (1971), 1–15.

111 Charnley, J., 'A Sterile Operating Theatre Enclosure', *British Journal of Surgery*, 51 (1964), 195–202, 202.

112 'Editorial: Airborne Infection in Surgery', *Journal of Hospital Infection*, 3 (1982), 215–16, 215.

113 See Nichols, R.L., 'Use of Prophylactic Antibiotics in Surgical Practice', *The American Journal of Medicine*, 70 (1981), 686–92.

114 New methods of laminar flow ventilation were used which the authors admitted might have had an effect on their results. Simchen, E., *et al.*, 'The Successful Use of Antibiotic Prophylaxis in Selected High-risk Surgical Patients Under Non-trial, Everyday Conditions', *Journal of Hospital Infection*, 1 (1980), 211.

115 Lidwell, O., 'Airborne Bacteria and Surgical Infection', *American Journal of Medicine*, 70 (1981), 693–7, 696.

116 Stinchfield, F.E., 'Editorial Comment: Total Hip Replacement', *Clinical Orthopaedics and Related Research*, 95 (1973), 3.

117 One article in *The Lancet* suggested that the best antibiotic for the prevention of infection was 'not known'. See *The Lancet* (4 March 1978), 482.

118 Pollard, J.P., *et al.*, 'Antibiotic Prophylaxis in Total Joint Replacement', *British Medical Journal*, 1 (17 March 1979), 707–9.

119 Charnley, J. and Eftekhar, N., 'Post-operative Infection in Total Prosthetic Replacement Arthroplasty of the Hip Joint', *Centre for Hip Surgery, Wrightington, Internal Publication*, 17 (1968), 1–17.

120 Waugh, *Charnley*, 153–67.

121 *British Medical Journal* (17 March 1979), 709.

122 Evrard, J., *et al.*, 'L'antibiotic Thérapie Préventive en Chirurgie Orthopédique', *Rev Chir. Orthop.*, 67, Supplement 2 (1981), 59.

123 See Charnley, J. 'Surgeon and Engineer', *The Lancet* (11 February, 1961), 325–6; and 'Surgeon and Engineer', *The Lancet* (25 February, 1961), 449.

124 The Stanmore, for instance, is manufactured by Biomet, but is not available in the United States. Interview: Dane A. Miller, Warsaw, 2002.

125 See Closhisy, John C. and Harris, William H., 'The Harris-Galante Uncemented Femoral Component in Primary Total Hip Replacement at 10 Years', *The Journal of Arthroplasty*, 14 (1999), 915–17; Lachiewicz, Paul F., *et al.*, 'A Prospective Study of 100 Consecutive Harris-Galante Porous Total Hip Arthroplasties', *The Journal of Arthroplasty*, 7 (1992), 519–26.

126 AAOS (2003), Hip Joint Replacement, 4–12.

6 Hips, Health Services and Quality

1 The NJR Annual Report 2005–2006, National Joint Registry (2006), 12. Some were paid for by the NHS, as we shall see below.

2 The data do not indicate what proportion of THRs used the Charnley prosthesis, but an indication is given by the fact that in 1968 around 6,000 Charnley prostheses were being produced annually. Given that the NHS total was 6,800 in 1968 it would appear that some 6 out of 6.8 or 88% were Charnley. However some of these may have been used outside the NHS or exported.

3 See Laing, William, *Hip Replacement and the NHS* (London: Office of Health Economics, 1982).

4 The best available data summarised show the number of THRs carried out in the NHS rising from under 7,000 in 1968 to just under 49,000 in 2005, a seven fold rise in less than four decades.

Although the data sources changed during the period, the data are comparable. Data were drawn from the Hospital In-Patient Enquiry scheme (HIPE) 1967 to 1985, and from Hospital Episode Statistics (HES) 1989–2005. HIPE data refer to Deaths and Discharges, and HES to Admissions which measure the same thing, that is persons admitted or equivalently, discharged. The use of admissions rather than consultant episodes avoids the problem of some patients having multiple episodes of care within a single admission. The graph shows a rise in 1967–1985 under HIPE followed by a gap to 1989 which is shown with a lower total than 1985. This is likely to be due to HES undercounting for the first few years. As shown the number rose sharply in 1992 to above the 1985 level and continued to rise sharply thereafter. The gap 1996–1998 is due to the non availability of HES online for those years.

5 Charnley, John, *Low Friction Arthroplasty of the Hip: Theory and practice* (New York: Springer-Verlag, 1979).

6 American Academy of Orthopedic Surgeons Data.

7 Laing, William, *Hip Replacement and the NHS* (1982).

8 Department of Health and Social Security, *Orthopaedic Services Waiting Times* (London: HMSO, 1981), 11 and 5.

9 British Orthopaedic Association, *Report to the Royal Commission on the National Health Service*, 1976–1977, 6. BS 6/1019.

10 Department of Health and Social Security, *Orthopaedic Services Waiting Times* (London: HMSO, 1981), 14.

11 Ferriman, Annabel, 'Year's Wait for 20% of Hip Operations', *The Times* (24 March 1982).

12 Williams, P.L., *et al.*, 'Pre-operative Medical Complications in Patients Admitted to Hospital for Elective Orthopaedic Surgery', *Health Trends*, 16 (1984), 66.

13 'Area Bias in Hip Surgery', *The Times* (12 June 1982).

14 See, Ham, Chris J., 'Power, Patients and Pluralism', in Barnard, Keith and Lee, Kenneth (eds), *Conflicts in the National Health Service* (London: Croom Helm, 1977), 99–120.

15 Discussion with Mr Christopher Faux, formerly of Preston hospital, 21/3/2007.

16 'Pensioner Demands Surgery Abroad', http://news.bbc.co.uk/1/hi/health/3162036.stm.

17 *Ibid.*

18 'Woman to sue for immediate treatment', http://news.bbc.co.uk/1/hi/health/2564657.stm.

19 Kay, Peter, 'Providing a Successful Hip Reconstruction Service: Ingredients, risks and opportunities', Paper presented at Wrightington Hospital, 2005.

20 Carvel, John, 'Row as Private Clinics Get NHS Staff', *The Guardian* (10 September 2003).

21 House of Commons, Health Select Committee, Fourth Report of 2004–5, on Independent Sector Treatment Centres.

22 For 1981 and 1986 THRs, see Williams *et al.*, in 'Total Hip Replacement', in Stevens, A. and Raftery, J. (eds) *Health Care Needs Assessment* (Oxford: Radcliffe Medical Press, 1995); for 1997/8, Williams *et al.*, 'Private Funding of Elective Hospital Treatments in England and Wales, for 1997–8,' *BMJ*, 320 (2003), 904–5; for 2003–5 National Joint Register Annual Report. Note that 2003 is for 9 months only. Data for 2003–5 reflect incomplete coverage of NJR.
23 BBC News Online, Thursday October 2, 2003. http://news.bbc.co.uk/1/hi/health/3153262.stm.
24 Consumers Union of U.S. Inc, 'Special Report: Keyhole surgery', *Consumer Reports on Health* (February 2002), 3.
25 *Calgary Herald* (23 January 2003).
26 Interview: Christopher Faux, Manchester, 2007.
27 Interview: Wayne Paprosky, Chicago, 2002.
28 Mulqueen, Éibhir, 'Putting an End to a Long and Painful Wait', *Irish Times* (9 March 2004).
29 See Porter, Michael E. and Teisberg, Elizabeth, *Redefining Heath Care: Creating value based competition on results* (Cambridge: Harvard Business School Press, 2006).
30 40% of medical expenditure in the United States is funded by the public sector. Krugman, Paul, 'Does Getting Old Cost Society Too Much?', *New York Times* (9 March 1997); Healy, William L., 'Economic Considerations in Total Hip Arthroplasty and Implant Standardization', *Clinical Orthopaedics and Related Research*, No. 311 (1995), 102; See also, Barber, T.C. and Healy, W.L., 'The Hospital Cost of Total Hip Arthroplasty: A Comparison Between 1981 and 1990', *Journal of Bone and Joint Surgery* 75A (1993), 321–5.
31 *The New York Times* (30 November 1971).
32 *The New York Times* (6 November 1977).
33 Gray, Bradford H., 'The Rise and Decline of the HMO: a chapter in the US health-policy history', in Stevens, Rosemary A., Rosenberg, Charles E. and Burns, Lawton R. (eds), *History and Health Policy in the United States* (New York: Rutgers University Press, 2006), 309–40.
34 Ziegenfuss, James T., *DRGs and Hospital Impact: An organizational systems analysis* (New York: McGraw-Hill Book Company, 1985), 2.
35 Somers, L.S., *et al.*, 'Clinician-Directed Hospital Cost Management for Total Hip Arthroplasty Patients', *Clinical Orthopaedics and Related Research*, 168 (1990), 168–75.
36 *The New York Times* (9 March 1997).
37 Somers, L.S., *et al.*, 'Clinician-Directed Hospital Cost Management' (1990), 168–75.
38 Cauley, Jane A., 'Epidemiology of Total Hip Replacement', *NIH Consensus Statement*, 19 (1994).
39 *The New York Times* (16 April 1997).
40 The average lifespan of a hip replacement varies, but ten years is considered a good outcome.
41 Levine, D., *et al.*, 'Cost Awareness and Cost Containment at the Hospital for Special Surgery', *Clinical Orthopaedics and Related Research*, No. 311 (1995), 121.

42 Proceedings and Reports of Councils and Associations, *Journal of Bone and Joint Surgery*, 47B (1965), 185.

43 http://arthritis.webmd.com/news/20031114/get-new-hip-walk-home-that-day (15 April 2004).

44 These include increased age, living alone, and the number of other health problems. Munin, Michael C., *et al.*, 'Predicting Discharge Outcome after Elective Hip and Knee Arthroplasty', *American Journal of Physical Medicine and Rehabilitation*, 74 (1995), 301.

45 Waddell, James P. and Morton, Jane, 'Generic Total Hip Arthroplasty', *Clinical Orthopaedics and Related Research*, 311 (1995), 115.

46 Fitzpatrick, R., *et al.*, 'Primary Total Hip Replacement Surgery: A systematic review of outcomes and modelling of cost-effectiveness associated with different prostheses', *Health Technology Assessment*, 2: 20 (1998), iv.

47 W.J. Maloney in an Expert Panel: Are costs limiting joint replacement? *Orthopedics Today*, 2006, 26 (2006) on Othosupersite.com (visited 17/3/07).

48 Source, OECD Health Data (CD) 2006, OECD Paris.

49 OECD figures; interpretations by James Raftery.

50 In one study, the rates of THR were higher for women over men, Caucasian over African American, and higher income over lower income. See Mahomed, Nizar, *et al.*, 'Rates and Outcomes of Primary and Revision Total Hip Replacement in the United States Medicare Population', *Journal of Bone and Joint Surgery*, 85A (2003), 27–32.

51 Source, OECD Health Data (CD) 2006. OECD Paris.
Health Statistics 2006 database. Data for a few other countries are available.

52 *Spending Review 2002*, 'Chapter 7: The Department of Health', HM Treasury, 2002, table 7.3.

53 Coventry, P.A. and Pickstone, J.V., 'From what and why did Genetics Emerge as a Medical Specialism in the 1970s in the UK? A case-history of research, policy and services in the Manchester Region of the National Health Service', *Social Sciences and Medicine*, 49 (1999), 1227–38.

54 Department of Health and Social Services, *Orthopaedic Services: Waiting times* (London: HMSO, 1981), 20.

55 Blume, Stuart, *Insight and Industry: On the dynamics of change in medicine* (Cambridge: MIT Press, 1992), 184–9.

56 Weinstein, M.C. & Stason, W.B. 'Foundations of Cost-effectiveness Analysis for Health and Medical Practices', *New England Journal of Medicine*, 296: 13 (1977), 716–21.

57 The earliest report in the literature was 'Cost Benefit/Cost Effectiveness of Medical Technologies: a case study of orthopaedic joint replacements' by Office of Technology Assessment (USA) in 1981 by Bentkover and Drew. Although this paper outlined the kind of data that would be required to carry out a cost effectiveness analysis, it did not perform such an analysis. See Williams, A., 'Economics of Coronary Artery Bypass Grafting', *British Medical Journal*, 291 (3 August 1985), 326–9.

58 Working as economic advisor to the Department of Health 1993–96, James Raftery attempted to locate the source of this estimate but failed. It seemed to be based on some very simple guesstimates.

59 *Effective Healthcare Bulletin: Total hip replacement*, Vol. 2, No. 7 (October 1995).
60 Dexter, Brian, 'Stellick Back in the Game with New, Pain-free Hip', *Toronto Star* (12 September 2002).
61 NICE, *Guidance on the Selection of Prostheses for Primary Total Hip Replacement* (London: HMSO, April 2000), 2.
62 Williams *et al.*, *Epidemiologically Based Needs Assessment: Total hip replacement* (1992), 16.
63 Wilcock, G.K., 'Benefits of Total Hip Replacement to Older Patients and the Community', *British Medical Journal* (1 July 1978), 39.
64 See Murray, D.W., Carr, A.J., Bulstrode, C.J., 'Which Primary Hip Replacement?' *Journal of Bone and Joint Surgery*, 77B (1994), 520–7.
65 Maloney, William J., in an Expert Panel: Are costs limiting joint replacement? *Orthopedics Today*, 2006, 26 (October 2006) on Othosupersite.com (visited 17/3/07).
66 Ezzet, Kace A., 'The Prevalence of Commercial Funding in Adult Lower Extremity Orthopaedic Research and its Impact on the Reporting of Scientific Results', Paper No. 031, *AAOS Annual Meeting Proceedings*, Vol. 4 (5–9 February 2003).
67 British Orthopaedic Association, *Total Hip Replacement: A guide to best practice* (1999), 6.
68 *Ibid.*
69 Hallett, J.P., 'Standards for Joint Replacement: A Clinician's Perspective', *Joint Replacement in the 1990s* (1992), 40.
70 Letter to J.T. Scales, National Orthopaedic Hospital, Stanmore from CE Phillips, Department of Scientific and Industrial Research, 28/07/55, National Archives, FD 23/1757.
71 Letter from H.J. Seddon, The Institute of Orthopaedics, Royal National Orthopaedic Hospital, Great Portland Street to Sir Landsborough Thompson, Medical Research Council, 13/10/55, National Archives, London, FD 23/1757.
72 *Ibid.*
73 British Standard, *Metallic Materials for Surgical Implants*, 1990, BS 726=52: Part 6: (1990).
74 Reynolds, L.A. and Tansey, E.M. (eds), *The Early Development of Total Hip Replacement. Wellcome Witnesses to Twentieth Century Medicine*, Volume 29 (London: The Wellcome Trust Centre for the History of Medicine at UCL, 2007).
75 US Congress, Medical Devices Hearings, *Hearings Before the Subcommittee on Public Health and Environment of the Committee on Interstate and Foreign Committee House of Representatives, Amendments to the Food, Drug and Cosmetic Act* (23–24 October 1973), 331. JLE 76–1358.
76 Abraham, John, 'The Science and Politics of Medicines Regulation', in Elston, Mary Ann (ed.), *The Sociology of Medical Science and Technology* (Oxford: Blackwells, 1997), 175.
77 Letter to Joe Barton from the Health Education and Human Services Division, US General Accounting Office, Health Education and Human Services Division, *Subcommittee on Oversight and Investigation* (September 24, 1998). GA1.13:HEHS-98–211.

78 US Congress, Medical Devices Hearings, *Hearings Before the Subcommittee on Public Health and Environment of the Committee on Interstate and Foreign Committee House of Representatives, Amendments to the Food, Drug and Cosmetic Act* (23–24 October 1973), 332. JLE 76–1358.

79 Hutt, Peter Barton, 'A History of Government Regulation of Adulteration and Misbranding of Medical Devices', *Food Drug Cosmetic Law Journal*, 44 (1995), 101–2.

80 Johnson, protecting the American Consumer, supra note 58, at 260, in Hutt, 'A History of Government Regulation' (1995), 107. And see the discussion of intra-ocular lenses in our Chapter 8.

81 Blume, Stuart, 'Medicine, Technology and Industry', in Cooter, Roger and Pickstone, John (eds), *Medicine in the Twentieth Century* (Amsterdam: Harwood Academic Publishers, 2000), 171–86.

82 Theodore Cooper was the Director of the National Heart and Lung Institution. The Committee was formed in 1969. Hutt, Peter Barton, 'A History of Government Regulation of Adulteration and Misbranding of Medical Devices' in Estrin, Norman F. (ed.), *The Medical Device Industry: Science, technology and regulation in a competitive environment* (New York: Marcel Dekker, 1990), 17–34, 25.

83 Hutt, 'A History of Government regulation' (1990), 25–6.

84 It was estimated by Rylee that it would cost between $10,000–$50,000 to develop an implant. Another company executive said it could cost as much as $150,000. US Congress, Medical Devices Hearings, *Hearings Before the Subcommittee on Public Health and Environment of the Committee on Interstate and Foreign Committee House of Representatives, Amendments to the Food, Drug and Cosmetic Act* (23–24 October 1973), 347. JLE 76–1358.

85 *Ibid.*, 333.

86 *Ibid.*, 341.

87 The Office of Technology Assessment, *Development of Medical Technology* (1976).

88 See Subcommittee on Oversight and Investigations, *Medical Device Regulation: The FDA's neglected child* (Washington: 1983).

89 Subcommittee on Oversight and Investigations, *Medical Device Regulation: The FDA's neglected child* (Washington: 1983), iv.

90 Munsey, Rodney M., 'Trends and Events in FDA Regulation of Medical Devices Over the Last Fifty Years', *Food and Drug Law Journal*, 50 (1995), 168.

91 Murray, D.W., *et al.*, 'Which Primary Hip Replacement?' *Journal of Bone and Joint Surgery*, 77B (1995), 526.

92 For the Exeter case, see Lee, A.J.C., 'Rough or Polished Surface on Femoral Anchorage Stems?' in Buchhorn, G.H., Willert, H.G. (eds), *Technical Principles, Design, and Safety of Joint Implants* (Toronto: Hogrefe and Hubber, 1994), 209–11; and Crawford, R.W., *et al.*, 'An 8–10-year Clinical Review of Comparing Matt and Polished Exeter Stems', *Orthopaedic Transactions*, 22 (1998), 40.

93 See Fisher J., *et al.*, 'Preliminary Study of the Effect of Ageing following Irradiation on the Wear of Ultrahigh-Molecular-Weight Polyethylene', *The Journal of Arthroplasty*, 10 (1995), 689–92; and Besong A.A., *et al.*,

'Quantitative Comparison of Wear Debris from UHMWPE that has and has not been Sterilised by Gamma Irradiation', *Journal of Bone and Joint Surgery*, 80B (1998), 340–4.

94 Sledge, Clement, B., 'Total Hip Replacement: 12 Years Later – What has Changed?', *NIH Consensus Statement* (1994), 17.

95 *National Institutes of Health Consensus Statement* (September 12–14, 1994), 25.

96 http://www.fda.gov/

97 Letter to James M. Jefferson, Committee on Labour and Human Resources US Senate from Bernice Steinhardt, Director, Health Services Quality and Public Health Issues, 3 June 1997, US General Accounting Office, *Medical Devices: FDA Review Times, 1989 Through 1996*, GA 1.13: HEHS- 97-146R.

98 *The New York Times* (18 September 1994).

99 *The New York Times* (30 June 1993).

100 *Ibid.*

101 *Ibid.*

102 US House of Representatives, *Hearing Before the Subcommittee on Technology of the Committee on Science* (2 November 1995), Y4. SCI2:104/26, 2.

103 *Ibid.*, 95–6.

104 Gordon, Stephen L., 'Federal Support for Research on Joint Replacements', in Morrey, B.F. (ed.), *Biological, Material and Mechanical Considerations of Joint Replacement* (New York: 1993), 393.

105 Office of Technology Assessment, *Medical Devices and the Veterans Administration: A technical memorandum*, Washington (February 1985). JLF 85–3045.

106 Gordon, Stephen L., 'Federal Support' (1993), 395.

107 Morris, Zoe, 'Fears Over Artificial Hips and Breasts', *The Evening Standard* (11 April 2002).

108 Sweetnam, D.R., 'A Surveillance Scheme with "Recommended List" of Artificial Joints', *Health Trends*, 13 (1981), 43–4.

109 See Malchau, H., *et al.*, 'Prognosis of Total Hip Replacement in Sweden: follow up of 92,675 Operation Performed 1978–1990', *Acta Orthopaedia Scand*, 64 (1993), 497–506.

110 Bulstrode, C.J.K., *et al.*, 'Designer Hips', *British Medical Journal*, 306 (20 March 1993), 732.

111 Reynolds and Tansey (eds), *Wellcome Witness Seminar.*

112 Harper and Gregg, 'Trent Regional Arthroplasty Study: Interim report and one year results', *Joint Replacement in the 1990s* (1992), 49.

113 Riordan, Pat, *et al.*, 'Lessons of a Hip Failure', *British Medical Journal*, 316 (27 June 1998), 1985.

114 Reynolds and Tansey (eds), *Wellcome Witness Seminar.*

115 Jefferys, D.B., 'The Regulation of Medical Devices and the Role of the Medical Devices Agency', *British Journal of Clinical Pharmacology*, 52 (2001), 229–35.

116 Jefferys, 'Regulation' (2001), 230.

117 Department of Health, *Manufacturer Registration Scheme: Guidance on process validation* (HMSO, London, 1981).

118 Supply/Health Service Industry Branch, 'The DHSS Manufacturer Registration Scheme', *Health Trends*, Vol. 18, No. 1, 22.

119 Jefferys, 'Regulation' (2001), 230–1.
120 Jefferys, 'Regulation' (2001), 230–1.
121 See their website at www.mhra.gov.uk
122 NICE, 'Hip Prostheses for Primary Hip Replacement', *Summary of Guidance Issued to the NHS in England and Wales*, Issue 4 (April 2002), 56.
123 National Audit Office. *Hip Replacements: an update*, HC956, 2002–3 (17 July 2003).
124 www.njrcentre.org.uk.
125 The NJR Annual Report 2005–2006 (National Joint Registry, 2006), 11.
126 Royal College of Surgeons of England, *The 3M Capital Hip System: The lessons learned from an investigation*, Leaflet (2001), 2.
127 See Massoud, *et al.*, 'Early Femoral Loosening in One Design of Cemented Hip Replacement', *Journal of Bone and Joint Surgery*, 79B (1997), 603–8.
128 The Royal College of Surgeons of England, *3M Capital Hip System* (2001), 13.
129 Harper, W.M., *et al.*, 'Register Exists in Trent Region: Letters', *British Medical Journal*, 316 (27 June 1998), 1985–1986.
130 CE marking is a declaration by the manufacturer that the product meets all the appropriate provisions of the relevant legislation implementing certain European Directives. CE marking gives companies easier access into the European market to sell their products without adaptation or rechecking. The initials 'CE' do not stand for any specific words but are a declaration by the manufacturer that his product meets the requirements of the applicable European Directives. Surgical Materials Testing Lab, News April 1998, http://www.smtl.co.uk/News/April98.html.
131 *Hansard* (21 April 1998).
132 Royal College of Surgeons, *The 3M Capital Hip System* (2001), 5.
133 Bonn, Dorothy, 'UK Patients with Hip Implants to be Reviewed', *The Lancet*, Vol. 351 (28 February 1998), 651.
134 Muirhead-Allwood, Sarah, 'Lessons of a Hip Failure', Editorial, *British Medical Journal*, Vol. 316 (28 February 1998), 644.
135 'A Thousand Faulty Hips in Hants and IoW', *Hampshire Chronicle* (27 February 1998).
136 The Royal College of Surgeons of England, *3M Capital Hip System*, 11–12.
137 Lawrence, Jeremy, 'Hip Implant Suffers Mass Failure Rate', *The Independent* (20 January 2003).
138 Lawrence, *The Independent* (20 January 2003).
139 *Financial Times* (10 May 2002).
140 *The Houston Chronicle* (5 February 2002).
141 There were 735 recalls of Sulzer hips in Quebec and that province had 'written confirmation' for millions of dollars. *The Montreal Gazette* (14 May 2002).

7 The Changing Nature of Patients: Expectations and Information

1 Cooper, William, *Scenes from Later Life* (London: Methuen, 1984), 222.
2 Anderson, Discussions, Emails and Phone Queries from Patients and their families, 2002–2004.
3 Interview with Professor Allan Gross, Chicago, 2002.

4 Report on British Orthopaedic Association Annual Meeting, *Journal of Bone and Joint Surgery*, 36B (1954), 507.
5 Denham, R.A. and Law, W. Alexander, 'Arthroplasty of the Hip', *Journal of Bone and Joint Surgery*, 39B (1957), 614.
6 Charnley, John, *Low Friction Arthroplasty of the Hip: Theory and practice* (New York: Springer-Verlag, 1979).
7 Cushner, F. and Friedman, R.J., 'Economic Impact of Total Hip Arthroplasty', *Southern Medical Journal*, 81 (1988), 1379–81.
8 Laing, William, *Hip Replacement and the NHS*, Office of Health Economics, London (1982), 3.
9 See Millar, Wayne J., 'Hip and Knee Replacement', *Statistics Canada Health Reports*, Vol. 14, No 1 (October 2002), 37–50.
10 Laucius, Joanne, 'Joint Replacement on Upswing: Active boomers demand new hips and knees', *Ottawa Citizen* (3 February 2003).
11 In 1997 it was estimated that one-third of all hip replacements were performed on patients younger than 65. *The New York Times* (16 April 1997).
12 Bulstrode, C., 'The Epidemiology of Total Hip Replacement', *Joint Replacement in the 1990s* (Bury St Edmunds, Mechanical Engineering Publications, 1992), 2.
13 See Hoagland, Steinbach, 'Primary Osteoarthritis of the Hip: Etiology and epidemiology', *Journal of the American Academy of Othopedic Surgeons*, 9 (September–October 2001), 320–7; Merx, H., *et al.*, 'International Variation in Hip Replacement Rates', *Annals of Rheumatic Disease*, 62 (2003), 222–6; Wendelboe, A.M., *et al.*, 'Relationships Between Body Mass Indices and Surgical Replacements of Knee and Hip Joints', *American Journal of Preventative Medicine*, 25 (2003), 290–5.
14 See Murray, R.O. and Duncan, C., 'Athletic Activity in Adolescence as an Aetiological Factor in Degenerative Hip Disease', *Journal of Bone and Joint Surgery*, 53B (1971), 406; 'Too Much Sport May Cause Hip Disease', *The Times* (8 September 1971).
15 Dunsmuir, Robert A., *et al.*, 'Increased Incidence of Primary Total Hip Replacement in Rural Communities', *British Medical Journal* (30 November 1996), 1370.
16 Northmore-Ball, M.D., 'Young Adults with Arthritic Hips', *British Medical Journal*, 315 (2 August 1997), 265.
17 Interview: Allan Gross, Chicago, 2002.
18 Interview: Wayne Paprosky, Chicago, 2002.
19 One of the biggest changes is the use of postoperative exercises and movement. Interview: William Harris, Boston, 2003.
20 Interview: Kevin Hardinge, Manchester, 2001.
21 Beinart, Jennifer and Sykes, M.K., *A History of the Nuffield Department of Anaesthetics, Oxford 1937–1987* (Oxford: Oxford University Press, 1987).
22 Brady, L.P., Enneking, W.F. and Franco, J.A., 'The Effect of Operating Room Environment on the Infection rate After Charnley Low-Friction Total Hip Replacement', *Journal of Bone and Joint Surgery*, 57A (1975), 81.
23 Pamphlet, *Minimal Incision Total Hip Arthroplasty*, 2003; Elsing, Debra Carr, 'Surgical Replacements are Becoming Less Invasive', *Capital Times* (10 December 2002).
24 These included golf, swimming laps, cycling, bowling, sailing and scuba diving. McGrory, Brian J., *et al.*, 'Participation in Sports after Hip and

Knee Arthroplasty: Review of literature and survey of surgeon preferences', *Mayo Clinic Proceedings*, Vol. 70 (April 1995), 346.
25 Pope, John, 'Hip Replacement Does Not Hinder Childbirth, Study Says', *The Times-Picayune* (6 February 2003).
26 Mallon, William J. and Callaghan, John J., 'Total Hip Arthroplasty in Active Golfers', *The Journal of Arthroplasty*, Vol. 7, Supplement (1992), 339–46.
27 Rothstein, Mervyn, 'The Peripatetic Sammy Davis', *New York Times* (30 May 1989).
28 Jackson was signed with the Chicago White Sox and his hip replacement surgery was used in an advertising campaign to sell tickets. *New York Times* (5 April 1993).
29 Crump, Thomas, *The Daily Telegraph Hip Replacement: What you really need to know* (London: Constable and Robinson, 2004), 133.
30 Silber, Irwin, *A Patient's Guide to Knee and Hip Replacement* (New York: Fireside, 1999), 225.
31 *British Medical Journal*, Vol. 285 (11 December 1982), 1710.
32 Frank Lewis, Director of Engineering at Richards quoted in the *New York Times* (20 November 1983).
33 Report on the Jephcott Lecture 'Medicine's Challenge to Engineering', *The Times* (16 June 1961).
34 *British Medical Journal* (7 October 1978), 1025.
35 'Researchers Use Robot for Hip Surgery on Dog', *New York Times* (8 May 1990).
36 L.C. Vaughan, 'Acrylic Femoral Head Prosthesis in the Cat', *The Veterinary Record* (22 October 1955), 804.
37 See the American Pet Products Manufacturers Association: http://www.appma.org/press_releasedetail.asp?id=84
38 Saxon, Wolfgang, 'Dr Howard A. Paul Dies at 59; Developed Robotic Surgical Arm', *New York Times* (13 February 1993).
39 Interview: Allan Gross, Chicago 2003.
40 Rothman, R.H. and Balderston, R.A., 'Patient-Surgeon Interview', in Booth, Robert E., *et al.*, *Total Hip Arthroplasty* (Philadelphia: Saunders, 1988), 58.
41 See Shepherd, Margaret M., 'Assessment of Function after Arthroplasty of the Hip', *Journal of Bone and Joint Surgery*, 36B (1954), 354–63.
42 See Andersson, Gunnar, 'Hip Assessment: A comparison of nine methods', *Journal of Bone and Joint Surgery*, 54B (1972), 621–5.
43 Lazansky, Mark, 'A Method for Grading Hips', *Journal of Bone and Joint Surgery*, 49B (1967), 644.
44 Kavanagh, B.F. and Fitzgerald, R.H., 'Clinical and Roentgenographic Assessment of Total Hip Arthroplasty: A new hip score', *Clinical Orthopaedics and Related Research*, 193 (1985), 133–40.
45 Field, R.E., *et al.*, 'The Oxford Hip Scores for Primary and Revision Hip Replacement', *Journal of Bone and Joint Surgery*, 87B (2005), 618–22; Pynsent, P.B., *et al.*, 'The Oxford Hip and Knee Outcome Questionnaires for Arthroplasty', *Journal of Bone and Joint Surgery*, 87B (2005), 241–8.
46 McDowell, W., *et al.*, 'A Method for Self-Assessment of Disability Before and After Hip Replacement Operations', *British Medical Journal* (23 September 1978), 858.
47 Birrell, F., *et al.*, 'Health Impact of Pain in the Hip Region with and without Radiographic Evidence of Osteoarthritis: A study of new attenders to

primary care – the PCR hip study group', *Annals of Rheumatic Diseases*, 59 (2000), 857–63.

48 For a discussion of the representation of the wheelchair see Iezzoni, Lisa, I., *When Walking Fails: Mobility problems of adults with chronic conditions* (Berkeley: University of California Press, 2003); and see Parr, Susie, Watson, Nick and Woods, Brian, 'Access, Agency and Normality: The wheelchair and the internet as mediators of disability', in Webster, Andrew, (ed.) *New Technologies in Health Care. Challenge and innovation* (Basingstoke: Palgrave, 2006), 161–76.

49 Rumbach, David, 'Plymouth Man Claims Damages for Failed Hip Prosthesis', *South Bend Tribune* (12 October 2003).

50 See Stewart, M.A. and Roter, D.L., 'Introduction', in Stewart, M.A. and Roter, D.L. (eds), *Communicating with Medical Patients* (Newbury Park: Sage, 1989).

51 See Lemaire, J., 'Informed Consent – a contemporary myth?', *Journal of Bone and Joint Surgery*, 88B (2006), 2–7; and Bhattacharyya, Timothy *et al.*, 'The Medical-Legal Aspects of Informed Consent in Orthopaedic Surgery', *Journal of Bone and Joint Surgery*, 87A (2005), 2395–400.

52 One of the largest UK suppliers of patient information (with regular updates on the quantitative risks of surgical procedures, including THR) is the private company EIDO Healthcare. It supplies over 200 hospitals and its information is given to about a third of UK patients undergoing surgery to assist with the consent process. More information can be found at http://www.eidohealthcare.com.

53 Video: *Total Hip Replacement: The new generation* (New York: Syracuse, 1986).

54 Patient Information Video: *Hip Replacement Surgery* (Wrightington: Wrightington Hospital NHS Trust, 1996).

55 Dunlop, Joyce, 'Medicine and the Media', *British Medical Journal* (14 February, 1987), 432.

56 For the BBC news article from the documentary see http://news.bbc.co.uk/1/hi/uk/57960.stm.

57 See Ley, P., *Communicating with Patients: Improving communication, satisfaction and compliance* (London: Chapman and Hall, 1988).

58 See Rhodes, Philip, 'Getting a New Hip Joint', *British Medical Journal*, Vol. 287 (10 September 1983), 747–8; Dunlop, Joyce, 'Getting Two New Hip Joints', *British Medical Journal*, Vol. 289 (December, 1984), 1750–1.

59 Anderson, Conversations with Patients, Chicago, September, 2002.

60 An example is Hardinge, Kevin, *Hip Replacement: The facts* (Oxford: Oxford Medical Publications, 1984).

61 Trahair, Richard, *All About Hip Replacement: A patient's guide* (Oxford: Oxford University Press, 1998); Allen, Ronald J., *et al.*, *Arthritis of the Hip and Knee: The active person's guide to taking charge* (Atlanta: Peachtree Publishers, 1998).

62 Crump, Thomas, *The Daily Telegraph Hip Replacement: What you really need to know* (London: Constable and Robinson, 2004); Silber, Irwin, *A Patient's Guide to Knee and Hip Replacement* (New York: Fireside, 1999).

63 Gabriel, Peggy, *Hip Replacement or Hip Resurfacing: A story of choices* (Bloomington, Indiana: AuthorHouse, 2003).

64 Klapper, Robert and Huey, Lynda, *Heal Your Hips: How to prevent hip surgery – and what to do if you need it* (New York: Wiley, 1999).

65 Grelsamer, Ronald P., *What Your Doctor May Not Tell You About Hip and Knee Replacement Surgery* (New York: Warner Books, 2004).
66 Margolies, Luisa, *My Mother's Hip: Lessons from the world of eldercare* (Philadelphia: Temple University Press, 2004).
67 *The Guardian* (17 March 1998).
68 *The Glasgow Herald* (21 November 1995).
69 *Daily Mirror* (17 November 1995).
70 *The Sunday Times Good Hospital Guide* (16 May 2004).
71 *Ibid.*
72 These walks for charity were established in the Northwest in Buxton in 1981. A walk around the Pavilion Gardens at Buxton was organised by the Arthritis and Rheumatism Research Council and attended by the Duke of Devonshire in 1987. 'Hips in Walk to Help Painful Joints', *Stockport Express Advertiser* (9 July 1987).
73 'The Walking Miracles!', *Prestwich Advertiser* (26 May 1988).
74 'Exclusive New Club Honours "Bionic Bill"', *Manchester Weekly News* (13 August 1988).
75 See http://disc.server.com/Indices/8340.html for Totallyhip and http://www.surfacehippy.info/ for Surfacehippy.
76 http://disc.server.com/Indices/8340.html for Totallyhip and http://www.surfacehippy.info/ for Surfacehippy.
77 *New York Times* (6 July 2005).
78 Tomes, Nancy, 'Patients or Health-Care Consumer? Why the history of contested terms matters', in Stevens, Rosemary, Rosenberg, Charles E. and Burns, Lawton (eds) *History and Health Policy in the United States: Putting the past back in* (New Jersey: Rutgers University Press, 2006), 83–110.
79 Arons, R.M. and Ruh, R.A., 'American Hospital Supply Company. An historic incubator of leadership talent', Korn/Ferry International at www.kornferry.com, see p. 5, 'It must be noted that American Hospital flourished and instituted these (rapid promotion) practices during the early 1970s when unusually unrestricted spending on heath care was prevalent..... Medical devices sold on a cost-plus basis and pricing pressures were few. And p6: 'New hires could earn over $50,000 their first year, a handsome sum in 1970.'... 'The go-go years came to an end with restrictions on reimbursements and other pricing pressures in the mid-80s.'
80 For more on the inflationary system and attempts at control, see Starr, Paul, *The Social Transformation of American Medicine: The rise of a sovereign profession and the making of a vast industry* (New York: Basic Books, 1982).

8 What is Past is Present

1 US Department of Health and Human Services. Center for Disease Control and Prevention: National Hospital Discharge Survey (www.cdc.gov/nchs/about/major/hdasd/nhdsdes.htm.
2 Platt had trained in Boston before WWII; see Butler, S.V.F., 'Academic Medicine in Manchester: The careers of Geoffrey Jefferson, Harry Platt and John Stopford', in *Bulletin of the John Rylands Library of the University of Manchester*, 87 (2007), forthcoming.

3 Waugh, *Charnley*, 115.
4 About two-thirds of the cemented stems are Exeter or Charnley and about 75% of the THR operations use cemented stems. National Joint Registry, 3rd Annual Report (2005–2006), 37, 42, 43.
5 From 1969 to his death, Charnley practised privately at the King Edward VII Hospital, Midhurst, West Sussex, see Waugh, *Charnley*, Chapter 15.
6 smith-nephew.com/new/item: 10 March 2005: Smith and Nephew Orthopaedics announces major promotions (21/03/2007).
7 See Chapter Three in Porter, M.E. and Teisberg, E.O., *Redefining Health Care: Creating value-based competition on results* (Boston: Harvard Business School Press, 2006), 71.
8 Angell, Marcia, *The Truth About the Drug Companies: How they deceive us and what to do about it* (New York: Random House, 2004).
9 Dane Miller in an Expert Panel: Are costs limiting joint replacement? Part Two, *Orthopedics Today*, 2006, Vol. 26; 18 (November 2006) on Orthosupersite.com (visited 17/3/07).
10 Jeffreys, D.B., 'The Regulation of Medical Devices and the Role of the Medical Devices Agency', *British Journal of Clinical Pharmacology*, Vol. 52 (2001), 229–35.
11 Miller, Dane, 'Orthopaedic Product Technology During the Second Half of the Twentieth Century', in Klenerman, Leslie (ed.), *The Evolution of Orthopaedic Surgery* (London: The Royal Society of Medicine Press, 2002), 223–4.
12 Dane Miller 'Are Costs Limiting Joint Replacement? Part Two' (November 2006) on Othosupersite.com (visited 17/3/07).
13 Metcalfe, J.S. and Pickstone, John, 'Replacing Hips and Lenses: Surgery, industry and innovation in post-war Britain', in Webster, Andrew (ed.), *New Technologies in Health Care: Challenge, change, innovation* (Basingstoke: Palgrave, 2006), 146–60.
14 www.rayner.com/history (visited 25 March 2007).
15 See Blume, Stuart, *Insight and Industry: On the dynamics of technological change in medicine* (Boston: MIT Press, 1992); Susskind, Charles, 'The Invention of Computed Tomography', *History of Technology*, Vol. 6 (London: Mansell Press, 1981); Kevles, B.H., 'The Physical Sciences and the Physician's Eye; Dissolving disciplinary boundaries', in *The Cambridge History of Science*, Vol. 5, Mary Jo Nye (ed.), *The Modern Physical and Mathematical Sciences* (Cambridge: Cambridge University Press, 2003), 615–33.
16 Blume, *Insight and Industry*, 186–9.
17 Pickstone, John V. (ed.), *Medical Innovations in Historical Perspective* (Basingstoke: Macmillan Palgrave, 1992), see 1–16 for discussion and references.
18 See Miller, 'Orthopaedic Product Technology' (2002), 211–28.
19 Schlich, Thomas, *Surgery, Science and Industry: A revolution in fracture care, 1950s–1990s* (Basingstoke: Palgrave, 2002).
20 See Reynolds, L.A. and Tansey, E.M. (eds), *The Early Development of Total Hip Replacement*, in *Wellcome Witnesses to Twentieth Century Medicine*, Volume 29 (London: The Wellcome Trust Centre for the History of Medicine at UCL, 2007).

21 For more on the need to engage medical professionals in reform, see Ham, Chris, 'Improving the Performance of Health Services: The role of clinical leadership', *The Lancet*, 361 (7 June 2003), 1978–80.
22 For a reassertion of the importance of academic medical centres in the US, see *Academic Health Centers: Leading change in the twentieth century*, Institute of Medicine (Washington DC: National Academies Press, 2004).
23 Smee, Clive T., *Speaking Truth to Power: Two decades of analysis in the Department of Health* (Oxford: Radcliffe Publications, 2005).
24 See Klein, Rudolph, *The New Politics of the NHS* (Oxford: Radcliffe, 2006).
25 The judgement of James Raftery.
26 Wallace, W. Angus, 'Independent Sector Treatment Centres: How the NHS is left to pick up the pieces', *British Medical Journal*, 332 (2006), 614.
27 House of Commons, Health Select Committee, Fourth Report of 2004–5, on Independent Sector Treatment Centres.
28 Ironically 'Centres of Clinical Excellence' is now the name of a private hospital company (*Financial Times*, 8 March 2007, article on a new hospital near Bath).
29 Wallace, BMJ letter (n. 26); and response to Stephen Black, Rapid responses, 19 March 2007 (www.bmj.com.cgi/eletters/332/7541/614, visited 05/03/07).
30 See Klein, *The New Politics* (2006).
31 See Porter and Teisberg, *Redefining HealthCare* (2006).
32 For privatisations, Pollock, Alyson, *NHS.plc* (London: Verso, 2004).
33 See Sarmiento, Augusto, *Bare Bones: A Surgeon's Tale: The price of success in American medicine* (Amherst: Prometheus, 2003).
34 The announcement of the subpoena was carried on the web sites of the major companies, and on orthopaedic business sites such as www.health-point.com/research/2005/04/04/ (visited 26/3/07).
35 Pickstone, John V., *Medicine and Industrial Society: A history of hospital development in Manchester and its region 1752–1946* (Manchester: Manchester University Press, 1985), passim and ch. 7.
36 See Butler, 'Academic Medicine' (2007).
37 Le Grand, Julian, *Motivation, Agency and Public Policy. Of knights, pawns and queens* (Oxford: Oxford University Press, 2003).

Bibliography

Archives

- National Archives, Kew, London.
  (Records of the Medical Research Council)
  FD 21/616
  FD 23/616
  FD 23/1757
  FD 23/1758
  BS6/1019
- US Federal Archives (Bethesda, MD) re. FDA etc.
- Charnley Papers, Medical Medical Archive, John Rylands University Library, University of Manchester
- Manufacturers' Catalogues etc., Thackray Museum, Leeds

Newspapers: see footnotes

Patient Information Booklets (in Date order)

Denham, R.A., *How To Be A Successful Hippie: Operative treatment for the painful hip by total joint replacement*, 1983.
Stepping Into the Future: Your guide to the pre-operative and postoperative management of total hip replacement, 1987.
Gross, Allan, Gross, Penny and Langer, Bernard (eds) *Surgery: A complete guide for patients and their families* (Toronto: HarperCollins, 1989).
Villar, Richard N., *Hip Replacement: A patient's guide to surgery and recovery* (London: Thorsons, 1995).
Trahair, Richard, *All About Hip Replacement: A patient's guide* (Oxford: Oxford University Press, 1998).
Klapper, Robert and Huey, Lynda, *Heal Your Hips: How to prevent hip surgery – and what to do if you need it* (New York: Wiley, 1999).
The British Medical Association, *Understanding Hip and Knee Arthritis Surgery*, 2003.
Grelsamer, Ronald P., *What Your Doctor May Not Tell You About Hip and Knee Replacement Surgery* (New York: Warner Books, 2004).
Wrightington, Wigan and Leigh, N.H.S., *Patient Information and Consent: Total hip replacement* (2004).
The Sunday Times Good Hospital Guide, 16 May 2004.

Professional and Official Reports

UK (in DATE order)

Medical Research Council, *War Memorandum*, No. 6 (1941).

Public Health Laboratory Service, 'Incidence of Surgical Wound Infection in England and Wales. A report', *The Lancet* (24 September 1960), 659–63.

Ministry of Health, *Hospital Building Note No. 26: The operating department* (London: HMSO, 1967).

Medical Research Council and Department of Health, *Report by Joint Working Party on Ventilation in Operation Suites* (London: MRC and DHSS, 1972).

Medical Research Council Trial Final Report, *Investigation of the Effect of Ultra-Clean Air in Operating Rooms on Surgical Sepsis* (London: MRC, 1979).

Medical Research Council, *Investigation of the Effect of Ultra-Clean Air in the Operating Room on Surgical Sepsis* (London: MRC, 1979).

Department of Health, *DH Manufacturer Registration Scheme: Guidance on process validation* (London: HMSO, 1981).

Department of Health and Social Security, *Orthopaedic Services Waiting Times* (London: HMSO, 1981).

Laing, William, *Hip Replacement and the NHS* (London: Office of Health Economics, 1982).

Department of Health, *The Control of Substances Hazardous to Health: Guidance for the initial assessment in hospitals* (London: HMSO, 1989).

British Orthopaedic Association, *Advisory Booklet on Consultant Orthopaedics and Trauma Services* (1990).

Department of Health Press Release, *The Capital Hip Replacement* (19 February 1998).

British Orthopaedic Association, *Total Hip Replacement: A guide to best practice* (1999).

NICE, *Guidance on the Selection of Prostheses for Primary Total Hip Replacement* (London: HMSO, April 2000).

The Royal College of Surgeons of England, *An Investigation of the Performance of the 3M Capital Hip System* (London: July 2001).

Committee on Patient Safety, Committee on Biological Implants, Tissue Work Group, *Musculoskeletal Allograft Tissue Safety Information Pamphlet*, 2003.

National Joint Registry for England and Wales, '1st Annual Report' (September 2004).

US (in DATE order)

US Congress, Medical Devices Hearings, *Hearings Before the Subcommittee on Public Health and Environment of the Committee on Interstate and Foreign Committee House of Representatives, Amendments to the Food, Drug and Cosmetic Act* (23–24 October 1973). JLE 76-1358.

Federal Supply Service, *Federal Standard Clean Room and Work Station Requirement, Controlled Environments*, No. 209B (August 1973).

Bentkover, Judith D. and Drew, Philip G., *Cost Benefit/Cost Effectiveness of Medical Technologies: A case study of orthopaedic joint implants*, Office of Health Technology Assessment Background Papers: Case Study #14 (Cambridge: Arthur D. Little, September 1981).

Report of the Subcommittee on Oversight and Investigations of the Committee on Energy and Commerce, US House of Representatives, *Medical Device Regulation: the FDA's neglected child: An oversight report on FDA implementation of the medical device amendments of 1976* (Washington: US Government Printing Office, 1983). Y4. EN 2/3:98-F.

Office of Technology Assessment, *Medical Devices and the Veterans Administration: A technical memorandum* (Washington: February 1985). JLF 85-3045.

General Accounting Office, *Medical Devices – Early warning of problems is hampered by severe under-reporting* (Washington, 1986).

Elelman, Karen B. (ed.), *New Medical Devices: Invention, development and use* (Washington: National Academy Press, 1988).

Lara, Maria Elena and Goodman, Clifford (eds), *National Priorities for the Assessment of Clinical Conditions and Medical Technologies: Report of a pilot study* (Washington: National Academy Press, 1990).

Geljns, Annetine C. (ed.), *Medical Innovation at the Crossroads: Modern methods of clinical investigation*, Vol. 1 (Washington: National Academy Press, 1990).

Geljns, Annetine C. and Halm, Ethan E. (eds), *Medical Innovation at the Crossroads: The changing economics of medical technology*, Vol. 2 (Washington: National Academy Press, 1991).

Geljns, Annetine C. (ed.), *Medical Innovation at the Crossroads: Technology and health care in an era of limits*, Vol. 3 (Washington: National Academy Press, 1992).

American Academy of Orthopaedic Surgeons, *Position Statement: Containing the cost of orthopaedic implants* (Rosemont: American Academy of Orthopaedic Surgeons, May 1992).

Geljns, Annetine C. and Dawkins, Holly V. (eds), *Medical Innovation at the Crossroads: Adopting a new medical technology*, Vol. 4 (Washington: National Academy Press, 1994).

National Institutes of Health Consensus Statement (12–14 September 1994). GA 1.13: HEHS-97-146R.

Rorabeck, Cecil H., 'Cost-Benefit Assessment of THR', *NIH Consensus Statement*, 1994, 27–9.

Rosenberg, Nathan, Geljns, Annetine C. and Dawkins, Holly V. (eds), *Medical Innovation at the Crossroads: Sources of medical technology*, Vol. 5 (Washington: National Academy Press, 1995).

Medical Technology Development and Commercialisation: *Hearing Before the Subcommittee on Technology of the Committee on Science*, US House of Representatives (2 November 1995), (Washington: US Government Printing Office, 1995). Y4. SCI2:104/26.

Medical Devices: FDA Review Times 1989 through 1996.

US General Accounting Office, 'Medical Device Regulation. Too early to assess European system's value as model for FDA' (Washington: GAO, 1996).

US General Accounting Office, Health Education and Human Services Division, *Subcommittee on Oversight and Investigation* (24 September 1998). GA1.13:HEHS-98-211.

General Accounting Office *Medical Technology: Implementing the good manufacturing practices regulation* (1998). GA1.5/2:T-PEMD-92-6.

Department of Health and Human Services, 'Guideline for Prevention of Surgical Site Infection', *Infection Control and Hospital Epidemiology*, Vol. 20, No. 1 (1999), 247–64.

Hanna, Kathi, *et al.* (eds), *Innovation and Invention in Medical Devices: Workshop summary* (Washington: National Academy Press, 2001).

AAOS and AATB, *What Can You Tell Me About Bone and Tissue Transplantation?*, Rosemont and Virginia (2002).

AAOS, Committee on Patient Safety, *Musculoskeletal Allograft Tissue Safety* Pamphlet (2003).

Sweden

Swedish Parliamentary Priorities Commission, *Priorities in Health Care* (1995).

Books

Allen, Ronald J., *et al.*, *Arthritis of the Hip and Knee: The active person's guide to taking charge* (Atlanta: Peachtree Publishers, 1998).
Angell, Marcia, *The Truth About the Drug Companies: How they deceive us and what to do about it* (New York: Random House, 2004).
Apley, Graham A., *Modern Trends in Orthopaedics* (London: Butterworths, 1972).
Audit Commission, *The Way to Go Home. Rehabilitation and remedial services for older people* (London: HMSO, 2000).
Ayliffe, G.A.J. and English, M.P., *Hospital Infection: From miasmas to MRSA* (Cambridge: Cambridge University Press, 2003).
Beinart, Jennifer and Sykes, M.K., *A History of the Nuffield Department of Anaesthetics, Oxford, 1937–1987* (Oxford: Oxford University Press, 1987).
Bijker, W.E., Hughes, T.P. and Pinch, T. (eds), *The Social Construction of Technological Systems: New directions in the sociology and history of technology* (Cambridge: MIT Press, 1989).
Bijker, W.E., *Of Bicycles, Bakelite and Bulbs: Toward a theory of sociotechnical change* (Cambridge: MIT Press, 1995).
Black, Jonathan and Dumbleton, John H. (eds) *Clinical Biomechanics. A case history approach* (New York: Churchill, 1981).
Blume, Stuart, *Insight and Industry: On the dynamics of change in medicine* (Cambridge: MIT Press, 1992).
Breneman, James C. with Dave Hager, *The Stryker Story: Homer's Iliad* (Kalamazoo: Phil Schubert and Associates, 1992).
Brown, Stanley A. and Lemmons, Jack E. (eds) *Medical Applications of Titanium and its Alloys: The materials and biological issues* (Pennsylvania: ASTM, 1996).
Bryder, Linda, *Below the Magic Mountain: A social history of tuberculosis in twentieth century Britain* (Oxford: Oxford University Press, 1988).
Cartwright, F.F., *Joseph Lister, the Man who Made Surgery Safe* (London: Weidenfield and Nicolson, 1963).
Catterall, A., *Orthopaedics* (Edinburgh: Churchill, 1992).
Charnley, John, *The Closed Treatment of Common Fractures* (London: Churchill Livingstone, 1950).
Charnley, John, *Compression Arthrodesis* (London: Churchill Livingstone, 1953).
Charnley, John, *Acrylic Cement in Orthopaedic Surgery* (Edinburgh: Livingstone, 1970).
Charnley, J., *Low Friction Arthroplasty of the Hip* (Berlin: Springer-Verlag, 1979).
Cooper, William, *Scenes from Later Life* (London: Methuen, 1984).
Cooter, R., *Surgery and Society in Peace and War: Orthopaedics and the organisation of modern medicine 1880–1948* (Basingstoke: Macmillan, 1993).

Cooter, Roger and Pickstone, John, *Companion to Medicine in the Twentieth Century* (London: Routledge, 2003).
Cox Gad, Shayne, *Safety Evaluation of Medical Devices* (New York: Marcel Dekker, 1997).
Crawford-Adams, J., *Outline of Orthopaedics*, 9th Edition (Edinburgh: Churchill Livingstone, 1981).
Crump, Thomas, *The Daily Telegraph Hip Replacement: What you really need to know* (London: Constable and Robinson, 2004).
Dhillon, B.S., *Medical Device Reliability and Associated Areas* (New York: CRC Press, 2000).
Ellis, David, *Technology and the Future of Health Care: Preparing for the next 30 years* (San Francisco & Chicago: Jossey-Bass Publishers & Health Forum Inc., 1999).
Estrin, Norman F., *The Medical Device Industry: Science, technology and regulation in a competitive environment* (New York: Marcel Dekker, 1990).
Finerman, Gerald, A.M., *et al.* (eds), *Total Hip Arthroplasty Outcomes* (Edinburgh: Churchill Livingstone, 1998).
Fisher, R.B., *Joseph Lister 1827–1912* (London: Macdonald and Jane's Publishers, 1977).
Furlong, Ronald, *Hydroxyapatite Ceramic: A decade of experience in hip arthroplasty* (London: Furlong Research Foundation, 1996).
Gabriel, Peggy, *Hip Replacement or Hip Resurfacing: A story of choices* (1st Books, 2003).
Ghista, Dhanjoo N., *Biomechanics of Medical Devices* (New York: Marcel Dekker, 1981).
Guttmann, Ludwig, *Spinal Cord Injuries: Comprehensive management and research* (Oxford: Blackwell Scientific Publications, 1973).
Hardinge, Kevin, *Hip Replacement: The facts* (Oxford: Oxford University Press, 1983).
Harris, William H. (ed.) *Advanced Concepts in Total Hip Replacement* (New Jersey: Slack, 1985).
Harrison, Bennett, *Lean and Mean: The changing landscape of corporate power in the age of flexibility* (New York: Basic Books, 1994).
Heck, Charles V., *1933–1983 Fifty Years of Progress* (Chicago: American Academy of Orthopaedic Surgeons, 1983).
Higson, Gordon R., *Medical Device Safety* (Bristol: Institute of Physics, 2001).
Hill, Alan G.S., *Modern Trends in Rheumatology* (London: Butterworths, 1971).
Hounshell, David A., *From the American System to Mass Production, 1800–1932* (Baltimore: Johns Hopkins University Press, 1988).
Howorth, F. Hugh, *Aeromicrobiologically Clean Zones for Surgery* (Howorth: Bolton, 1984).
Howell, Joel, *Technology in the Hospital: Transforming patient care in the twentieth century* (Baltimore: Johns Hopkins University Press, 1995).
Hubbard, Richard F. and Rodengen, Jeffrey L., *Biomet: From Warsaw to the world* (Fort Lauderdale: Write Stuff Enterprises, 2002).
Hughes, Thomas P., *Human-Built World: How to think about technology and culture* (Chicago: University of Chicago Press, 2004).
Hughes, Thomas, P., *Networks of Power: Electrification in Western society 1880–1930* (Baltimore: Johns Hopkins University Press, 1983).

Iezzoni, Lisa I., *When Walking Fails: Mobility problems of adults with chronic conditions* (Berkeley: University of California Press, 2003).
Institute of Medicine, *Academic Health Centres: Leading change in the twentieth century* (Washington DC: National Academies Press, 2004).
Jeffrey, Kirk, *Machines in Our Hearts: The cardiac pacemaker, the implantable defibrillator and American health care* (Baltimore: Johns Hopkins University Press, 2001).
Keith, Arthur, *Menders of the Maimed* (Huntington NY: Robert E. Krieger Publishing Company, 1975).
Klein, Rudolph, *The New Politics of the NHS* (Oxford: Radcliffe, 2006).
Klenerman, Leslie (ed.) *The Evolution of Orthopaedic Surgery* (London: The Royal Society of Medicine Press, Ltd, 2002).
LeFanu, James, *The Rise and Fall of Modern Medicine* (London: Little, Brown and Company, 1999).
Le Grand, Julian, *Motivation, Agency and Public Policy. Of knights, pawns and queens* (Oxford: Oxford University Press, 2003).
Lawrence, G. (ed.) *Technologies of Modern Medicine: Proceedings of a seminar held at the Science Museum* (London: Science Museum, 1994).
Lettin, Alan, *Was It Something I Said?* (London: Memoir Club, 2005).
Lerner, Barron H., *The Breast Cancer Wars: Hope, fear and the pursuit of a cure in twentieth century America* (Oxford: Oxford University Press, 1988).
Lester, Richard K. and Piore, Michael J., *Innovation: The missing dimension* (Cambridge: Harvard University Press, 2004).
Ley, P., *Communicating With Patients: Improving communication, satisfaction and compliance* (London: Chapman and Hall, 1988).
Lowy, I. (ed.) *Medicine and Change: Historical and sociological studies of medical innovation* (Paris: INSERM/John Libbey, 1993).
MacKennzie, Donald and Wajcman, Judy (eds) *The Social Shaping of Technology: How the refrigerator got its hum* (Buckingham: Open University Press, 1985).
Margolies, Luisa, *My Mother's Hip: Lessons from the world of eldercare* (Philadelphia: Temple University Press, 2004).
McKee, G.K., *The McKee Method of Total Hip Replacement* (London: Deloro, 1973).
Meiksins, Peter and Smith, Chris, *Engineering Labour: Technical workers in comparative perspective* (London: Verso, 1996).
Metcalfe, J.S., *Evolutionary Economics and Creative Destruction* (London: Routledge, 1998).
Mohanna, Kay and Ruth Chambers, *Risk Matters in Healthcare: Communicating, explaining and managing risk* (Abingdon: Radcliffe Medical Press, 2001).
Moynihan, Ray and Alan Cassels, *Selling Sickness* (New York: Nation Books, 2005).
Nye, David E., *Narratives and Spaces: Technology and the construction of American culture* (Devon: University of Exeter Press, 1997).
O'Mahony, Patrick, *Nature, Risk and Responsibility: Discourses of biotechnology* (New York: Routledge, 1999).
Older, John (ed.) *Implant Bone Interface* (London: Springer-Verlag, 1990).
Olson, W.K., *The Litigation Explosion* (New York: Truman Tally, 1991).
Paprosky, Wayne G. (ed.), *Revision Total Hip Arthroplasty* (Rosemont IL: American Academy of Orthopaedic Surgeons, 2001).
Petroski, Henry, *Pushing the Limits: New adventures in engineering* (New York: Vintage Books, 2004).

Pickstone, John V., *Medicine and Industrial Society: A history of hospital development in Manchester and its region, 1752–1946* (Manchester: Manchester University Press, 1985).
Pickstone, John V. (ed.) *Medical Innovations in Historical Perspective* (Basingstoke: Macmillan, 1992).
Pickstone, John V., *Ways of Knowing* (Manchester: Manchester University Press, 2000).
Pollock, Alyson, *NHS.plc* (London: Verso, 2004).
Porter, Michael E. and Teisberg, Elizabeth, *Redefining Health Care: Creating value based competition on results* (Cambridge: Harvard Business School Press, 2006).
Postak, Paul D., *Highly Cross-Linked Polyethylene Modular Acetabular Designs: Performance characteristics* (Ohio: Orthopaedics Research Laboratories, 2003).
Postel, M., *et al.*, *Total Hip Replacement* (Berlin: Springer-Verlag, 1987).
Rang, Mercer, *The Story of Orthopaedics* (Philadelphia: Saunders, 2000).
Reynolds, L.A. and Tansey, E.M. (eds) *The Early Development of Total Hip Replacement*, in *Wellcome Witnesses to Twentieth Century Medicine*, Volume 29 (London: The Wellcome Trust Centre for the History of Medicine at UCL, 2007).
Rose, W. and Carless, A., *A Manual of Surgery*, 7th edition (London: Balliere and Co, 1920).
Rosenberg, Nathan, *Exploring the Black Box: Technology, economics and history* (Cambridge: Cambridge University Press, 1994).
Rowe, Carter R., *Lest We Forget: Orthopaedics at the Massachusetts General Hospital and Ward 1, 1900–1996* (Dublin NH: William L. Bauhan, 1996).
Sarmiento, Augusto, *Bares Bones: A surgeon's tale: The price of success in American medicine* (Amherst: Prometheus Books, 2003).
Schlich, Thomas, *Surgery, Science and Industry: A revolution in fracture care, 1950s–1990s* (Basingstoke: Palgrave, 2002).
Schwartz-Cowan, Ruth, *More Work for Mother: The ironies of household technologies from the open hearth of the microwave* (New York: Basic Books, 1983).
Sedel, L. and Cabanela, M.E. (eds) *Hip Surgery: Materials and developments* (London: Martin Dunitz, 1998).
Silber, Irwin, *A Patient's Guide to Knee and Hip Replacemen*t (New York: Fireside, 1999).
Simpson (ed.) *Modern Trends in Biomechanics* (London: Butterworths, 1970).
Sivash, K.M., *Alloplasty of the Hip Joint* (Moscow: Central Institute for Traumatology and Orthopaedics, 1967).
Smee, Clive T., *Speaking Truth to Power: Two decades of analysis in the Department of Health* (Oxford: Radcliffe Publications, 2005).
St-Clair Strange, F.G., *The Hip* (London: William Heinemann Medical Books, 1965).
Stanton, Jennifer (ed.), *Innovations in Health and Medicine: Diffusion and resistance in the twentieth century* (London: Routledge, 2002).
Starr, Paul, *The Social Transformation of American Medicine. The rise of a sovereign profession and the making of a vast industry* (New York: Basic Books, 1982).
Stevens, A. and Raftery, J. (eds) *Health Care Needs Assessment* (Oxford: Radcliffe Medical Press, 1995)
Stevens, Rosemary A., Rosenberg, Charles E. and Burns, Lawton R. (eds) *History and Health Policy in the United States* (New York: Rutgers University Press, 2006).

Swinburn, W.R., *Wrightington Hospital 1933–1983: The story of the first 50 years* (Wrightington: Wrightington Hospital, 1983).
Tenner, Edward, *Our Own Devices: The past and future of body technology* (New York: Alfred A. Knopf, 2003).
Timmermann, Carsten and Anderson, Julie (eds) *Devices and Designs: Medical technologies in historical perspective* (Basingstoke: Palgrave, 2006).
Wainwright, Penny, *Opposite the Infirmary: A history of the Thackray company 1902–1990* (Leeds: Thackray Medical Museum, 1997).
Walker, Peter S., *Human Joints and their Artificial Replacement* (Springfield: Charles C. Thomas, 1977).
Waugh, William, *John Charnley: The man and the hip* (London: Springer-Verlag, 1990).
Waugh, William, *A History of the British Orthopaedic Association: The first seventy-five years* (London: British Orthopaedic Association, 1993).
Webster, Andrew (ed.) *New Technologies in Health Care: Challenge, change, innovation* (Basingstoke: Palgrave, 2006).
Williams, M. *et al.*, *Total Hip replacement: Epidemiologically Based Needs Assessment* (Health Care Evaluation Unit, University of Bristol, 2004).
Willmann, G. and Zweymuller, K., *Bioceramics in Hip Joint Replacement: Proceedings of the 5th ceramic technology symposium* (New York: Thieme Medical Publishers, 2000).
Worboys, M., *Spreading Germs: Disease theories and medical practice in Britain 1865–1900* (Cambridge: Cambridge University Press, 2000).
Ziegenfuss, James T., *DRGs and Hospital Impact: An organizational systems analysis* (New York: McGraw-Hill Book Company, 1985).

Selected Primary and Secondary Articles

Abraham, John and Reed, Tim, 'Trading Risks for Markets: The international harmonisation of pharmaceuticals regulation', *Health Risk and Society*, 3 (2001), 113–28.
Abraham, John, 'The Science and Politics of Medicines Regulation', in Elston, Mary Ann (ed.), *The Sociology of Medical Science and Technology* (Oxford: Blackwell, 1997), 153–82.
Amstutz, Harlan C. and Grigoris, Peter, 'Metal on Metal Bearings in Hip Arthroplasty', *Clinical Orthopaedics and Related Research*, No. 329S (1996), S11–S34.
Anderson, Julie, 'Greenhouses and Bodysuits: The challenge to knowledge in early hip replacement surgery 1960–1982', in Timmermann, Carsten and Anderson, Julie (eds), *Devices and Designs: Medical technologies in historical perspective* (Basingstoke: Palgrave Macmillan, 2006), 175–92.
Anderson, Julie, 'Innovation and Locality: Hip replacement in Manchester and the North West of England', *Bulletin of the John Rylands Library of the University of Manchester*, 87 (2007), forthcoming.
Andersson, Gummar, 'Hip Assessment: A comparison of nine methods', *Journal of Bone and Joint Surgery*, 54B (1972), 621–5.
Arons, R.M. and Ruh, R.A., 'American Hospital Supply Company. An historic incubator of leadership talent', Korn/Ferry International at www.kornferry.com.

Barber, T.C. and Healy, W.L., 'The Hospital Cost of Total Hip Arthroplasty: A comparison between 1981 and 1990', *Journal of Bone and Joint Surgery*, 75A (1993), 321–5.

Battacharyya, Timothy, *et al.*, 'The Medical-Legal Aspects of Informed Consent in Orthopaedic Surgery', *Journal of Bone and Joint Surgery*, 87A (2005), 2395–400.

Bentley, G. and Duthi, R.B., A Comparative Review of the McKee-Farrar and Charnley Total Hip Prostheses, *Clinical Orthopaedics*, 95 (1973), 127–42.

Billingham, L.J., *et al.*, 'Methods for the Analysis of Quality-of-life and Survival Data in Health Technology Assessment', *Health Technology Assessment*, Vol. 3, No. 10 (1999).

Black, J., 'Metal on Metal Bearings: A practical alternative to metal on polyethylene total joints', *Clinical Orthopaedics and Related Research*, Supplement, No. 329 (1996), 244–55.

Blume, Stuart S., 'The Rhetoric and Counter-rhetoric of a "Bionic" Technology', *Science, Technology and Human Values*, 22 (1997), 31–56.

Blume, Stuart, 'Medicine, Technology and Industry', in Cooter, Roger and Pickstone, John (eds), *Medicine in the Twentieth Century* (Amsterdam: Harwood Academic Publishers, 2000), 171–86.

Bobyn, J. Dennis, 'THR: Femoral Fixation', *NIH Consensus Statement*, 1994, 43–7.

Boersma, Kees, 'Tensions Within an Industrial Research Laboratory: The Philips Laboratory's X-ray department between the wars', *Enterprise and Society*, 4 (2003), 65–98.

Boettcher, W.G., 'Total Hip Arthroplasties in the Elderly: Morbidity, mortality and cost effectiveness', *Clinical Orthopaedics and Related Research*, 274 (1992), 30–4.

Bowling, Ann, *et al.*, 'Let's Ask Them: A national survey of definitions of quality of life and its enhancement among people aged 65 and over', *International Journal of Aging and Human Development*, 56 (2003), 269–306.

Boyan, Barbara D. and Bolander, Mark E., 'Techniques and Models', in Morrey, B.F. (ed.), *Biological, Material and Mechanical Considerations of Joint Replacement* (New York: Raven, 1993), 451–2.

Brady, L.P., Enneking, W.F. and Franco, J.A., 'The Effect of Operating Room Environment on the Infection Rate after Charnley Low-Friction Total Hip Replacement', *Journal of Bone and Joint Surgery*, 57A (1975), 80–3.

Braley, Silas A., 'Acceptable Plastic Implants', in Simpson (ed.), *Modern Trends in Biomechanics* (1970), 25–51.

Brock, Lord, 'The Importance of Environmental Conditions, especially Temperature in the Operating Room and Intensive Care Ward', *British Journal of Surgery*, 62 (1975), 253–8.

Bucholz, H.W., *et al.*, Management of Deep Infection of Total Hip Replacement', *Journal of Bone and Joint Surgery*, 63B (1981), 342–53.

Bulstrode, C.J.K., *et al.*, 'Designer Hips: Don't let your patient become a fashion victim', *British Medical Journal*, 306 (20 March 1993), 732–3.

Butler, S.V.F., 'Academic Medicine in Manchester: The careers of Geoffrey Jefferson, Harry Platt and John Stopford', in *Bulletin of the John Rylands Library of the University of Manchester*, 87 (2007), forthcoming.

Callaghan, John and Savory, Carlton G., 'The Uncemented Porous-Coated Anatomic Total Hip Prosthesis', *Journal of Bone and Joint Surgery*, 70A (1988), 337–46.

Cantor, David, 'Cortisone and the Politics of Drama', in Pickstone, J.V. (ed.), *Medical Innovations in Historical Perspective* (Basingstoke: Macmillan, 1992), 165–84.

Cantor, David, 'Representing the "Public": Medicine, charity and emotion in twentieth century Britain', in Sturdy, Steve (ed.), *Medicine, Health and the Public Sphere in Britain* (London: Routledge, 2002), 145–68.

Carroll-Burke, Patrick, 'Tools, Instruments and Engines: Getting a handle n the specificity of engine science', *Social Studies of Science*, 31 (2001), 593–625.

Cauley, Jane A., 'Epidemiology of Total Hip Replacement', *NIH Consensus Statement* (1994), 19–25.

Charnley, John, 'A Sterile Operating Theatre Enclosure', *British Journal of Surgery*, 51 (1964), 195–202.

Charnley, John, 'The Bonding of Prostheses to Bone by Cement', *The Journal of Bone and Joint Surgery*, 46B (1964), 518–29.

Charnley, John, 'Operating-Room Conditions', Letter to the Editor, *The Lancet* 286 (1965), 907–8.

Charnley, John, 'A Biomechanical Analysis of the Use of Cement to Anchor the Femoral Head Prosthesis', *The Journal of Bone and Joint Surgery*, 47B (1965), 354–63.

Charnley, John, 'Limitations of the Germ Theory', *The Lancet* (13 July 1968), 101–2.

Charnley, John and Eftekhar, N., 'Post-operative Infection in Total Prosthetic Arthroplasty of the Hip', *British Journal of Surgery*, 56(9) (1969), 641–9.

Charnley, John, 'Postoperative Infection after Total Hip Replacement with special reference to Air Contamination in the Operating Room', *Centre for Hip Surgery, Wrightington Hospital Internal Publication*, No. 38 (March 1972), 1–24.

Charnley, John, 'The Organisation of a Special Hip Centre for Hip Surgery', *British Journal of Hospital Medicine*, 9 (March 1973), 389–96.

Charnley, John, 'The Development of the Centre for Hip Surgery at Wrightington Hospital', in Swinburn, W.R. *Wrightington Hospital 1933–1983: the story of the first 50 years* (1983), 236–42.

Christel, Pascal, 'Ceramics for Joint Replacement', in Morrey, B.F. (ed.), *Biological, Material and Mechanical Considerations of Joint Replacement*, New York (1993), 303.

Clohisy, John C. and Harris, William H., 'The Harris-Galante Uncemented Femoral Component in Primary Total Hip Replacement at 10 Years', *Journal of Arthroplasty*, 14 (1999), 915–17.

Collier, John P., *et al.*, 'The Tradeoffs Associated with Modular Hip Prostheses', *Clinical Orthopaedics and Related Research*, 311 (1995), 91–101.

Convery, F. Richard, *et al.*, 'The Relative Safety of Polymethylmethacrylate', *The Journal of Bone and Joint Surgery*, 57A (1975), 57–64.

Cooter, Roger, 'The Politics of a Spatial Innovation: Fracture clinics in inter-war Britain', in Pickstone, John V. (ed.), *Medical Innovations in Historical Perspective* (Basingstoke: Macmillan, 1992), 146–64.

Cooter, Roger, and Pickstone, John, 'From Dispensary to Hospital: Medicine, community and workplace in Ancoats, 1828–1948', *Manchester Regional History Review*, 7 (1993), 73–84.

Coventry, Mark B., 'Lessons Learned in 30 Years of Total Hip Arthroplasty', *Clinical Orthopaedics and Related Research*, 274 (1982), 22–9.

Coventry, P.A. and Pickstone, J.V., 'From What and Why did Genetics Emerge as a Medical Specialism in the 1970s in the UK? A case-history of research, policy and services in the Manchester Region of the National Health Service', *Social Sciences and Medicine*, 49 (1999), 1227–38.

Cushner, F. and Friedman, R.J., 'Economic Impact of Total Hip Arthroplasty', *Southern Medical Journal*, 81 (1988), 1379–81.

Duff-Barclay, I., Scales, J.T. and Wilson, J.N., 'The Development of the Stanmore Total Hip Replacement', *Proceedings of the Royal Society of Medicine*, 59 (1966), 948–51.

Dunlop, Joyce, 'Getting Two New Hip Joints', *British Medical Journal*, 289 (22–29 December, 1984), 1750–1.

Dunlop, Joyce, 'Medicine and the Media', *British Medical Journal* (14 February, 1987), 432–3.

Dunsmuir, Robert A., *et al.*, 'Increased Incidence of Primary Total Hip Replacement in Rural Communities', *British Medical Journal* (30 November 1996), 1370.

Elson, R.A., *et al.*, 'Antibiotic-loaded Acrylic Cement', *Journal of Bone and Joint Surgery*, Vol. 59B, 1977, 200–5 and *British Medical Journal* (24 May 1980), 1241.

Evans, Brian G., *et al.*, 'The Rationale for Cemented Total Hip Arthroplasty', *Orthopedic Clinics of North America*, 24 (1993), 559–609.

Ezzet, Kace A., 'The Prevalence of Commercial Funding in Adult Lower Extremity Orthopaedic Research and its Impact on the Reporting of Scientific Results', Paper No. 031, *AAOS Annual Meeting Proceedings*, Vol. 4 (5–9 February 2003).

Falkner, Alex, 'Casing the Joint: The material development of artificial hips', in Ott, Katherine, *et al.* (eds), *Artificial Parts, Practical Lives: Modern histories of prosthetics* (New York: New York University Press, 2002), 199–226.

Fender, D., *et al.*, 'Need for a National Arthroplasty Register: Letters', *British Medical Journal*, 313 (19 October 1996), 1007.

Field, R.E., *et al.*, 'The Oxford Hip Scores for Primary and Revision Hip Replacement', *Journal of Bone and Joint Surgery*, 87B (2005), 618–22.

Fitzpatrick, R., *et al.*, 'Primary Total Hip Replacement Surgery: A systematic review of outcomes and modelling of cost-effectiveness associated with different prostheses', *Health Technology Assessment*, Vol. 2, No. 20 (1998).

Fowler, J.L., *et al.*, 'Experience with the Exeter Total Hip Replacement since 1970', *Orthopedic Clinics of North America*, Vol. 19, No. 3 (July 1988), 477–89.

Gelman, *Solutions for Prevention of Airborne Infections* (Italy: Gelman, n.d.).

Giacomini, Mita K., 'The Which-Hunt: Assembling health technologies for assessment and rationing', *Journal of Health Politics, Policy and Law*, 24 (1999), 715–58.

Gillespie, W.J., *et al.*, 'The Incidence of Cancer following Total Hip Replacement', *Journal of Bone and Joint Surgery*, 70B (1988), 539–42.

Ginsburg, Gill, 'Human Factors Engineering: A tool for medical device evaluation in hospital procurement decision making', *Journal of Biomedical Informatics*, 38 (2005), 213–19.

Golan, Tal, 'The Emergence of the Silent Witness: The legal and medical reception of X-rays in the USA', *Social Studies of Science*, 34 (2004), 469–99.

Gordon, Stephen L., 'Federal Support for Research on Joint Replacements', in Morrey, B.F. (ed.) *Biological, Material and Mechanical Considerations of Joint Replacement* (New York: Raven Press, 1993), 393–5.

Gosden, P.E., 'Importance of Air Quality and Related Factors in the Prevention of Infection in Orthopaedic Implant Surgery', *The Journal of Hospital Infection*, 39:3 (1998), 173–80.

Gray, Bradford H., 'The Rise and Decline of the HMO: A chapter in US health policy history', in Stevens, Rosemary A., Rosenberg, Charles E. and Burns, Lawton R. (eds) *History and Health Policy in the United States* (New York: Rutgers University Press, 2006), 309–40.

Greene, Jeremy A., 'Attention to "Details": Etiquette and the pharmaceutical salesman in postwar America', *Social Studies of Science*, 34 (2004), 192–271.

Haboush, E., 'A New Operation for Arthroplasty of the Hip Based on Biomechanics, Photoelasticity, Fast Setting Dental Acrylic and Other Considerations', *Bulletin of Hospital Joint Disease*, 14 (1953), 242–77.

Ham, Chris J., 'Power, Patients and Pluralism', in Barnard, Keith and Lee, Kenneth (eds), *Conflicts in the National Health Service* (London: Croom Helm, 1977), 99–120.

Ham, Chris, 'Improving the Performance of Health Services: The role of clinical leadership', *The Lancet*, 361 (7 June 2003), 1978–80.

Hargreaves, Sally, 'It's Hip to Be Smooth', *The Lancet*, 356 (25 November 2000), 1910.

Harris, W.H., 'A New Total Hip Implant', *Clinical Orthopaedics and Related Research*, 81 (1971), 105–13.

Harris, W.H., 'The First 32 Years of Total Hip Arthroplasty', *Clinical Orthopaedics and Related Research*, 274 (January 1992), 6–11.

Harrold, A.J., 'Arthritis of the Hip', *British Medical Journal* (21 June 1969), 741–6.

Harrold, A.J., 'Outlook for Hip Replacement', *British Medical Journal*, 284 (16 January 1982), 139.

Harwood, Jonathan, 'Engineering Education between Science and Practice: Rethinking the historiography', *History and Technology*, 22 (2006), 53–79.

Healy, William L., 'Economic Considerations in Total Hip Arthroplasty and Implant Standardization', *Clinical Orthopaedics and Related Research*, 311 (February 1995), 102–8.

Hesket, John, 'Industrial Design', in Boris Ford (ed.) *Modern Britain: The Cambridge Cultural History*, Vol. 9 (Cambridge: CUP, 1992).

Hiderfjäll, Patrik, 'Biotronik: 40 years of German entrepreneurship in medical technology', in Timmermann, Carsten and Anderson, Julie (eds) *Devices and Designs: Technology and medicine in historical perspective* (Basingstoke: Palgrave, 2006), 112–36.

Holmberg, Sven, 'Life Expectancy after Total Hip Arthroplasty', *Journal of Arthroplasty*, 7 (June 1992), 183–6.

Howe, Chester W., 'Bacterial Flora of Clean Wounds and Its Relation to Subsequent Sepsis', *American Journal of Surgery*, 107 (May 1964), 696–700.

Howie, J.W., 'The Public Health Laboratory Service', *The Lancet* (6 March 1965), 501–5.

Huo, Michael H. and Salvati, Eduardo A., 'Hip Wear: Clinical aspects', in Morrey, B.F. (ed.), *Biological, Material and Mechanical Considerations of Joint Replacement* (New York: Raven Press, 1993).

Howorth, F.H., 'Prevention of Airborne Infection during Surgery', *American Society of Heating, Refrigerating and Air-Conditioning Engineers, Transactions*, 91 (1985), 291.

Hungerford, David S. and Jones, Lynne C., 'The Rationale for Cementless Total Hip Replacement', *Orthopedic Clinics of North America*, 24 (1993), 617–26.

Hutt, Peter Barton, 'A History of Government Regulation and Misbranding of Medical Devices', *Food, Drug, Cosmetic Law Journal*, 44 (1989), 101–2.

Hutt, Peter Barton, 'A History of Government Regulation of Adulteration and Misbranding of Medical Devices', in Estrin Norman F., *The Medical Device Industry: Science, technology and regulation in a competitive environment* (New York: Marcel Dekker, 1990), 17–34.

Illingworth, Charles, 'On the Interdependence of Science and the Healing Art', *Annals of the Royal College of Surgeons*, No. 35 (1964), 1–14.

Jacobs, J.H., 'Osteoarthritis of the Hip Joint', *Annals of Physical Medicine*, 5 (1960), 318–26.

Jeffreys, D.B., 'The Regulation of Medical Devices and the Role of the Medical Devices Agency', *British Journal of Clinical Pharmacology*, 52 (2001), 229–35.

Jones, Judy, 'National Register Will Monitor Hip Replacements', *British Medical Journal*, 320 (29 April 2000), 1163.

Jones, L.C. and Hungerford, D.S., 'Cement Disease', *Clinical Orthopaedics and Related Research*, 225 (1987), 192–206.

Joyce, Kelly, 'Appealing Images: Magnetic resonance imaging and the production of authoritative knowledge', *Social Studies of Science*, 35 (2005), 437–62.

Judet, J. and Judet, R., 'The Use of an Artificial Femoral Head for Arthroplasty of the Hip Joint', *Journal of Bone Joint Surgery*, 32B (1950), 166–73.

Kavanagh, B.F. and Fitzgerald, R.H., 'Clinical and Roentgenographic Assessment of Total Hip Arthroplasty: A new hip score', *Clinical Orthopaedics and Related Research*, 193 (1985), 133–40.

Kevles, B.H. 'The Physical Sciences and the Physician's Eye: Dissolving disciplinary boundaries', in *The Cambridge History of Science*, Vol. 5, Nye, Mary Jo (ed.), *The Modern Physical and Mathematical Sciences* (Cambridge: Cambridge University Press, 2003), 615–33.

Klenerman, Leslie, 'Arthroplasty of the Hip', in Klenerman, Leslie (ed.), *The Evolution of Orthopaedic Surgery* (London: Royal Society of Medicine Press, 2002), 13–23.

Lamoreaux, Naomi R., *et al.*, 'Beyond Markets and Hierarchies: Toward a new synthesis of American business history', *The American Historical Review*, 108 (2003), 404–33.

Lazansky, Mark, 'A Method for Grading Hips', *Journal of Bone and Joint Surgery*, 49B (1967), 644–51.

Learmonth, I.D., 'Prevention of Infection in the 1990s', *Orthopaedic Clinics of North America*, 24 (1993), 735–41.

Lehoux, Pascale and Blume, Stuart, 'Technology Assessment and the Sociopolitics of Health Technologies', *Journal of Health Politics, Policy and Law*, 25 (2000), 1083–120.

Lemaire, J., 'Informed Consent – A contemporary myth?', *Journal of Bone and Joint Surgery*, 88B (2006), 2–7.

Lemons, Jack E., 'Keynote Address: Biomaterials for Total Joint Replacements', in Morrey, B.F. (ed.), *Biological, Material and Mechanical Considerations of Joint Replacement* (New York: Raven Press, 1993), 13–25.

Lerouge, Sophie, *et al.*, 'Alumina Ceramic in total joint replacement', in Sedel, L. and Cabenela, M.E. (eds) *Hip Surgery: Materials and developments* (London: Martin Dunitz, 1998).

Levine, David B., Cole, Brian J. and Rodeo, Scott, 'Cost Awareness and Cost Containment at the Hospital for Special Surgery', *Clinical Orthopaedics and Related Research*, 311 (February 1995), 117–24.

Lidwell, O.M., 'Sepsis after Total Hip or Knee Joint Replacement in Relation to Airborne Contamination', *Philosophical Transactions of the Royal Society London*, B302 (1983), 583–92.

Ling, R.S.M., 'Cementing Technique in the Femur', *Techniques in Orthopaedics*, 6 (1991), 34–9.

Ling, R.S.M., 'The Use of a Collar and Precoating on Cemented Femoral Stems Is Unnecessary and Detrimental', *Clinical Orthopaedics and Related Research*, 285 (December 1992), 73–83.

Ling, Robin, 'The Development of the Exeter hip', in Faux, J. (ed.), *After Charnley* (Preston: The John Charnley Trust, 2002), 146–69.

Liscomb, Paul R., 'Reconstructive Surgery for Bilateral Hip-Joint Disease in the Adult', *Journal of Bone and Joint Surgery*, 47A (1965), 1–29.

Lister, J., 'On a New Method of Treating Compound Fractures, Abscesses etc., with observations on the conditions of suppuration', *The Lancet*, 89 (1867), 326–9.

Lister, J., 'On the Antiseptic Principle in the Practice of Surgery', *British Medical Journal*, 2 (1867), 246–8.

Livingston, B.J., *et al.*, 'Complications of Total Hip Arthroplasty Associated with the Use of an Acetabular Component with a Hylamer Liner', *Journal of Bone and Joint Surgery*, 79A (1997), 1529–38.

Lord, Frederick C., *et al.*, 'Infection in Bone Allografts. Incidence, Nature and Treatment', *The Journal of Bone and Joint Surgery*, 70A (1988), 369–76.

Lowry, Stella, 'Norwich Hip', *British Medical Journal*, Vol. 297 (July 1988), 247.

Madhok, R., *et al.*, 'Utilization of Upper Limb Joint Replacements During 1972–90: The Mayo Clinic experience', *Proceedings of the Institution for Mechanical Engineers*, 207 (1993), 239–44.

Malchau, H., *et al.*, 'Prognosis of Total Hip Replacement in Sweden: follow up of 92,675 Operations Performed 1978–1990', *Acta Orthopaedia Scand*, 64 (1993), 497–506.

Mansfield, Edwin, 'Industrial Innovation in Japan and the United States', *Science* (30 September 1988), 1769–74.

Marcus, Randall E., *et al.*, 'Don't Throw Away the Austin Moore', *The Journal of Arthroplasty*, 7 (1992), 31–6.

Marks, H.M., 'Medical Technologies: Social contexts and consequences', in Bynum, W. and Porter, R. (eds) *Companion Encyclopaedia of the History of Medicine*, Vol. 2 (London: Routledge, 1993), 1592–618.

Marlowe, Donald E., 'Regulation of Biomaterials and Medical Devices', in Morrey, B.F. (ed.), *Biological, Material and Mechanical Considerations of Joint Replacement* (New York: Raven Press Ltd, 1993), 397–404.

Mayor, Michael B., 'Discussion of Regulatory Issues', in Morrey, B.F., *Biological, Material and Mechanical Considerations of Joint Replacement* (New York: Raven Press Ltd., 1993), 423–4.

McDowell W., *et al.*, 'A Method for Self-Assessment of Disability before and after Hip Replacement Operations', *British Medical Journal* (September 23, 1978), 858.

McKee, G.K. and Chen, S.C., 'The Statistics of the McKee-Farrar Method of Total Hip Replacement', *Clinical Orthopaedics and Related Research*, 95 (September 1973), 26–33.

McKee, G.K., 'Faith Healing by a Fellow of the Royal College of Surgeons', *The Mancroft Review*, 1–3.

McKee, G.K., 'Developments in Total Hip Joint Replacement', *Proceedings of the Institution of Mechanical Engineers*, 181 (1966), 85–9.

McKee, G.K., 'Development of Total Prosthetic Replacement of the Hip', *Clinical Orthopaedics and Related Research*, 72 (1970), 85–103.

Melkerson, Mark N., *et al.*, 'Regulatory Perspective for Orthopaedic Devices', in Morrey, B.F., *Biological, Material and Mechanical Considerations of Joint Replacement* (New York: Raven Press Ltd, 1993), 405–22.

Merx, H., *et al.*, 'International Variation in Hip Replacement Rates', *Annals of Rheumatic Disease*, 62 (2003), 222–6.

Metcalfe, J.S. and Pickstone, John, 'Replacing Hips and Lenses: Surgery, industry and innovation in post-war Britain', in Webster, Andrew (ed.), *New Technologies in Health Care: Challenge, change, innovation* (Basingstoke: Palgrave, 2006), 146–60.

Michelson, James D. and Riley, Lee H., 'Considerations in the Comparison of Cemented and Cementless Total Hip Prostheses', *Journal of Arthroplasty*, 4 (1989), 327–34.

Millar, Wayne J., 'Hip and Knee Replacement', *Statistics Canada Health Reports*, 14 (2002), 37–50.

Miller, Dane A., 'Orthopaedic Product Technology During the Second Half of the Twentieth Century', in Klenerman Leslie (ed.) *The Evolution of Orthopaedic Surgery* (London: RSM Press, 2002), 211–25.

Mirowski, Philip and Van Horn, Robert, 'The Contract Research Organisation and the Commercialisation of Scientific Research', *Social Studies of Science*, 35 (2005), 503–48.

Moore, A.T. and Bohlman, H.R., 'The Classic Metal Hip Joint: A case report by Austin T. Moore and Harold R. Bohlman – 1943', *Clinical Orthopaedics* (1983), 3–6.

Moore, A.T., 'Metal Hip Joint: A new self-locking Vitallium prosthesis', *Southern Medical Journal*, 45 (1952), 1015–19.

Morris, Richard W., 'Designer Hips', *British Medical Journal*, 306 (24 April 1993), 1124.

Mowatt, G., *et al.*, 'When and How to Assess Fast-changing Technologies: A comparative study of medical application of four generic technologies', *Health Technology Assessment*, Vol. 1, No. 14 (1997).

Müller, Maurice E., 'Lessons of 30 Years of Total Hip Arthroplasty', *Clinical Orthopaedics and Related Research*, 274 (January 1982), 12–21.

Müller, Maurice M., 'The Benefits of Metal-on-Metal Total Hip Replacements', *Clinical Orthopaedics and Related Research*, 311 (1995), 54–9.

Munsey, Rodney R., 'Trends and Events in FDA Regulation of Medical Devices over the Last Fifty Years', *Food and Drug Law Journal*, 50 (1995), 163–77.

Murray, D.W., *et al.*, 'Which Primary Hip Replacement?', *Journal of Bone and Joint Surgery*, 77B (1995), 520–7.

Nixon, John, 'Failure Patterns after Total Hip Replacement', *British Medical Journal*, 286 (January 1983), 166–70.

Nixon, John, 'Non-cemented Hip Prostheses', *British Medical Journal*, 290 (June 1985), 1680–1.

Ong, L.M.L., *et al.*, 'Doctor-Patient Communication: A review of the literature', *Social Science and Medicine*, 40 (1995), 903–18.

Oppenheimer, B.S., Oppenheimer, E.T., Stout, A.P. and Eirich, F.R., 'Further Studies of Polymers as Carcinogenic Agents in Animals', *Cancer Research*, 15 (1955), 333.

Pascal, Christel, 'Ceramics for Joint Replacement', in Morrey, B.F. (ed.), *Biological, Materials and Mechanical Considerations of Joint Replacement* (New York: Raven, 1993).

Pickstone, John V., 'Bones in Lancashire: Towards long-term contextual analysis of medical technology', in Timmermann, Carsten and Anderson, Julie (eds) *Devices and Designs: Technology and medicine in historical perspective* (Basingstoke: Palgrave, 2006), 17–36.

Pickstone, John V., 'Contested Cumulations: Configurations of cancer treatments through the twentieth century', *Bulletin of the History of Medicine*, 81 (2007), 164–96.

Pickstone, John V., 'Science and Technology in Manchester: An introduction to the history', in John V. Pickstone (ed.), 'The History of Science and Technology in the North West', *Manchester Region History Review*, 18 (2007), 1–18.

Pickstone, John V., 'Medicine in Manchester, Manchester in Medicine, 1750–2005', in Butler, Stella V.F. and Pickstone, John V. (ed.), special number on Manchester Medicine, *John Rylands Library Bulletin*, 87 (2007), forthcoming.

Pinch, Trevor J. and Bijker, Wiebe E., 'The Social Construction of Facts and Artefacts: Or how the sociology of science and the sociology of technology might benefit each other', *Social Studies of Science*, 14 (1984), 399–441.

Powell, J.N., *et al.*, 'Cardiac Arrest Associated with Bone Cement', *British Medical Journal* (8 August 1970).

Pynsent, P.B., *et al.*, 'The Oxford Hip and Knee Outcome Questionnaires for Arthroplasty', *Journal of Bone and Joint Surgery*, 87B (2005), 241–8.

Rasmussen, Nicolas, 'The Moral Economy of the Drug Company – Medical Scientist Collaboration in Interwar America', *Social Studies of Science*, 34 (2004), 161–85.

Rhodes, Philip, 'Getting a New Hip Joint', *British Medical Journal*, 287 (10 September 1983), 747–8.

Ring, P.A. 'Total Replacement of the Hip Joint. A review of a thousand operations', *Journal of Bone and Joint Surgery*, 56B (1974), 44–58.

Ring, P.A., 'Long Term Results of the Uncemented Hip Replacement', in Noble, A. and Galasko, C. (Manchester: Manchester University Press, 1987), 214–21.

Ring, P.A., 'Press-Fit Prostheses: Clinical experience', in Reynolds, D. and Freeman, M. (eds), *Osteoarthritis in the Young Adult Hip Options for Surgical Management* (Edinburgh: Churchill Livingstone, 1989), 210–32.

Riordan, Pat, *et al.*, 'Lessons of a Hip Failure', *British Medical Journal*, 316 (27 June 1985), 644.

Ritter, M.A., *et al.*, 'The Use of a Hip Score for Evaluation of the Results of Total Hip Arthroplasty', *The Journal of Arthroplasty*, 5 (1990), 187–9.

Ritter, M.A. and Campbell, Elaine D., 'Long-term Comparison of the Charnley, Müller, and Trapezoidal-28 Total Hip Prostheses', *The Journal of Arthroplasty*, 2 (1987), 299–308.

Royal College of Surgeons of England, 'Report on Design of Operating Theatre Suites', *Annals of the Royal College of Surgeons of England*, 34 (1964), 217–88.

Sage, William M., Unfinished Business: 'How Litigation Relates to Health Care Regulation', *Journal of Health Politics, Policy and Law*, 28 (2003), 387–419.

Saleh, K.J., *et al.*, 'Economic Evaluations in the Hip Arthroplasty Literature: Lessons to be learned', *Journal of Arthroplasty*, 14 (1999), 527–32.

Scales, John T. and Wilson, J.N., 'Some Aspects of the Development of the Stanmore Total Joint Prosthesis', *Reconstructive Surgery and Traumatology*, 11 (1969), 20–39.

Shepherd, Margaret M., 'Assessment of Function after Arthroplasty of the Hip', *Journal of Bone and Joint Surgery*, 36B (1954), 354–63.

Sledge, Clement B., 'Total Hip Replacement: 12 Years Later – What has Changed?', *NIH Consensus Statement* (1994), 17–18.

Smith-Petersen, M.N., 'Arthroplasty of the Hip: A new method', *Journal of Bone and Joint Surgery*, 21 (1939), 269–88.

Sochart, David H., *et al.*, 'Joint Responsibility: The need for a national arthroplasty register', *British Medical Journal*, 313 (13 July 1996), 66–7.

Solomon, Michael I., *et al.*, 'Survivorship of Cemented Total Hip Arthroplasty in Patients 50 Years of Age or Younger', *The Journal of Arthroplasty*, 7, Supplement (1992), 347–52.

Somers, L.S., *et al.*, 'Clinician-directed Hospital Cost Management for Total Hip Arthroplasty Patients', *Clinical Orthopaedics*, 168 (1990), 168–75.

Stanton, Jennifer, 'Making Sense of Technologies in Medicine', *Social History of Medicine*, 12 (1999), 437–48.

Stinchfield, F.E., 'Editorial Comment: Total hip replacement', *Clinical Orthopaedics and Related Research*, 95 (1973), 3.

Sturdy, Steve, 'From the Trenches to the Hospitals at Home: Physiologists, clinicians and oxygen therapy', in Pickstone, J.V. (ed.) *Medical Innovations in Historical Perspective* (Basingstoke: Macmillan, 1992), 104–23.

Susskind, Charles, 'The Invention of Computed Tomography', *History of Technology*, Vol. 6 (London: Mansell Press, 1981), 39–80.

Supply/Health Service Industry Branch, 'The DHSS Manufacturer Registration Scheme', *Health Trends*, Vol. 18, No. 1, 22.

Swanson, S.A.V., 'Design and Testing of Replacement Prostheses', *Proceedings of the Royal Society of London A*, 192 (1976), 173–89.

Sweetnam, D.R., 'A Surveillance Scheme with "Recommended List" of Artificial Joints', *Health Trends*, 13 (1981), 43–4.

Taylor, D.G., 'The Costs of Arthritis and the Benefits of Joint Replacement Surgery', *Proceedings of the Royal Society of London A*, 192 (1976), 145–55.

Thompson, F.R., 'Vitallium intramedullary hip prosthesis – preliminary report', *New York State Journal Medicine*, 52:30 (1952), 11–20.

Tronzo, Raymond G., *et al.*, 'Elevation of Intramedullary Pressure when Methylmethacrylate is Inserted in Total Hip Arthroplasty', *The Journal of Bone and Joint Surgery*, 56A (1974), 714–18.

Tomes, Nancy, 'Patients or Health-Care Consumer? Why the History of Contested Terms Matters', in Stevens, Rosemary, Rosenberg, Charles E. and Burns, Lawton (eds) *History and Health Policy in the United States: Putting the past back in* (New Jersey: Rutgers University Press, 2006), 83–110.

Vaughan, Diane, 'The Role of Organization in the Production of Techno-Scientific Knowledge', *Social Studies of Science*, 29 (1999), 913–43.

Waddell, James P. and Morton, Jane, 'Generic Total Hip Arthroplasty', *Clinical Orthopaedics and Related Research*, 311 (1995), 109–16.

Wallace, W. Angus, 'Independent Sector Treatment Centres: How the NHS is left to pick up the pieces', *British Medical Journal*, 332 (2006), 614.

Wangensteen, Owen H. and Wangensteen, Sarah D., 'Military Surgeons and Surgery, Old and New: An instructive chapter in management of contaminated wounds', *Surgery*, 62 (1967), 1102–24.

Weinstein, M.C. and Stason, W.B., 'Foundations of Cost-Effectiveness Analysis for Health and Medical Practices', *New England Journal of Medicine*, 296 (1977), 716–21.

Wendelboe, A.M., *et al.*, 'Relationships between Body Mass Indices and Surgical Replacements of Knee and Hip Joints', *American Journal of Preventative Medicine*, 25 (2003), 290–5.

Wilcock, G.K., 'Benefits of Total Hip Replacement to Older Patients and the Community', *British Medical Journal* (1 July 1978), 39.

Wiles, Philip, 'The Surgery of the Osteo-Arthritic Hip', *British Journal of Surgery*, 45 (1958), 488–97.

Wise, Jacqui, 'Older, Cheaper Hip Implants are Best', *British Medical Journal*, 313 (2 November 1996), 1103.

Wood, P.H.N., 'Arthritis Calling for Surgical Treatment – Aetiology and Epidemiology', *Proceedings of the Royal Society of London B*, 192 (1976), 131–43.

Yamamuro, T., 'Zirconia Ceramic for the Femoral Head of a Hip Prosthesis', in Sedel, L. and Cabanela, M.E. (eds) *Hip Surgery: Materials and developments* (London: Martin Dunitz, 1998), 41.

Videos

Bone Allografts for Surgical Implantation (1994).
Cementless Revision Total Hip Replacement (1988).
Extended Femoral Osteotomy (1993).
From Thackray With Care (1989).
Hip Replacement (1989).
Hip Surgery: Recovering at Home (1994).
Modern Ceramics (2000).
Total Hip Replacement: Improving Quality of Life (2001).
Total Hip Replacement: The New Generation (1986).
Understanding Total Hip Replacement (1989).

Websites

AOA National Joint Replacement Registry
http://www.dmac.adelaide.edu.au/ aoanjrr/aoanjrr.jsp?section=history

BBC News Online, Thursday 2 October 2003.
http://news.bbc.co.uk/1/hi/health/3153262.stm

BBC News Online, 'Pensioner Demands Surgery Abroad'.
http://news.bbc.co.uk/1/hi/health/3162036.stm

BBC News Online, 'Goo measure could help hip ops', 22 October 2003.
http://news.bbc.co.uk/1/hi/health/3180436.stm

Biomedical Market Newsletter, 7 August 2003.
http://mailiwant.com/links.jsp?linkid=7622&subid=852197&campid=11498

Biomedical Market Newsletter, 31 July 2003.
http://mailiwant.com/links.jsp?linkik=7544&subid=852197&campid=11419

Canadian Institute for Health Information
http://secture.cihi.ca/cihiweb/dispPage.jsp?cw_page=AR30_2004sum_e

DePuy
www.depuy.com

FDA, 'Medical Devices Recalls of Zirconia Ceramic Femoral Heads for Implants', September 2001
http://www.fda.gov/cdrh/recalls/zirconiahip.html

National Joint Registry Summary Report
http://www.njrcentre.org.uk

Expert Panel
Orthosupersite.com

Korn/Ferry International
http://www.kornferry.com

Rayner company history
www.rayner.com/history
Smith and Nephew
http://www.smithandnephew.com

http://www.smtl.co.uk/News/April98.html

Which? Online, Medical Devices
http://sub.which.net/health/reports/apr2002he10t13/

Index